ENGINES
COMPACT EQUIPMENT

FUNDAMENTALS
OF SERVICE **FOS**

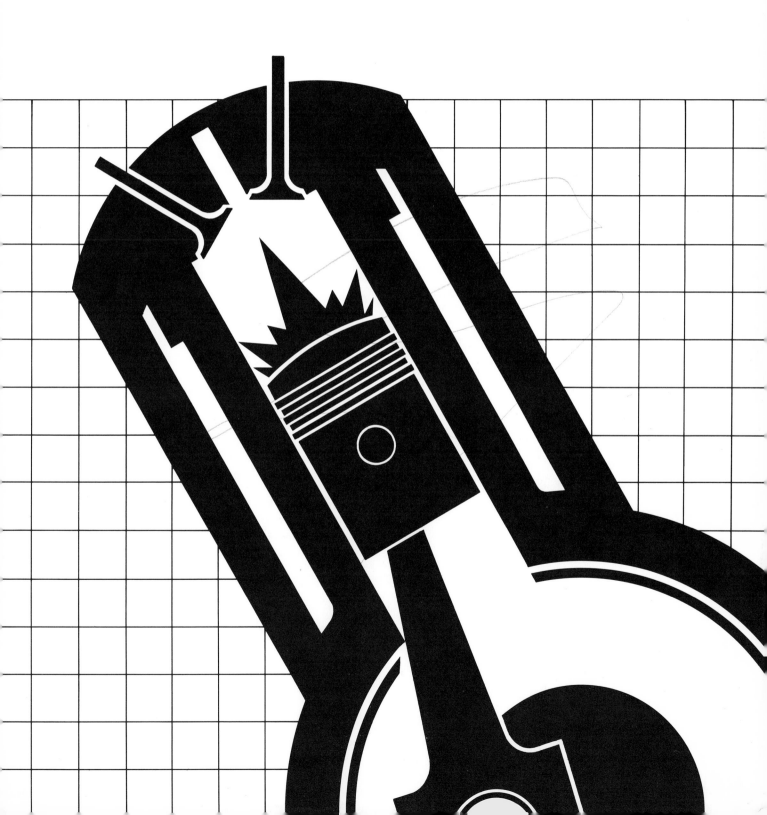

PUBLISHER

Fundamentals of Service (FOS) is a series of manuals created by Deere & Company. Each book in the series was conceived, researched, outlined, edited, and published by Deere & Company. Authors were selected to provide a basic technical manuscript which could be edited and rewritten, if necessary, by staff editors.

PUBLISHER: JOHN DEERE TECHNICAL SERVICES, Dept. F, John Deere Road, Moline, Illinois 61265; Manager: John A. Conrads; Service Training Manager: John D. Spuller.

FUNDAMENTAL WRITING SERVICE EDITORIAL STAFF
Managing Editor: Louis R. Hathaway
Editor: Michael H. Johanning
Editor: Laurence T. Hammond
Promotions: Annette M. LaCour

TO THE READER

The purpose of this manual is to help you gain greater understanding of the test and repair procedures for engines commonly found on compact equipment.

By understanding the material covered in this manual, along with hands-on experience, you should have the basic knowledge and skills required of a beginning technician in a compact equipment service department.

COVERAGE IN THIS MANUAL

Topics covered in this manual are limited to engines found on equipment rated up to 40 PTO horsepower (30 kW). Coverage includes both gasoline (2- and 4-cycle) and diesel engines below that power limit.

The engines discussed in this manual may be found on the following equipment: chain saws, weed trimmers, lawn mowers, riding mowers, lawn and garden tractors, compact utility tractors, skid-steer loaders, and others.

To understand the test and repair procedures, it is helpful to understand how the engine works. The first four chapters of this manual cover the components and operation of engines and fuel systems. Chapters 5 through 10 discuss test and repair procedures.

OTHER MANUALS IN THE SERIES

The FOS series covering compact equipment systems consists of four manuals. The three topics besides **Engines** are:

- **Electrical Systems**
- **Hydraulics**
- **Power Trains**

Each manual is backed up by a Teacher's Guide, Student Workbook and a set of 35 mm color slides.

FOR MORE INFORMATION

Write for a free catalog "Teaching Materials from John Deere." The catalog describes each of the subjects in the compact equipment series and gives ordering instructions. The catalog also describes other subjects in the John Deere series of training materials for agricultural and industrial applications. Send your request for the catalog to:

John Deere Service Training
Dept. F
John Deere Road
Moline, IL 61265

ACKNOWLEDGEMENTS

The editorial staff wishes to express its gratitude to the following people who contributed significantly to the development of these books:

Mark Lindsley — Author
Donald E. Borgman — Field Services Supervisor, John Deere Horicon Works
Philip L. Lane — Division Service Manager, John Deere Company, Minneapolis
Thomas L. Shelton — Consumer Products Service Manager, John Deere Company, East Moline

We would also like to thank the service departments from the following John Deere units for the time many individuals devoted to the review of this manuscript: John Deere Company, East Moline; John Deere Company, Portland; John Deere Company, Syracuse; John Deere Horicon Works; John Deere Tractor Works.

Thanks also to Marson Rinkenberger of the Minnesota Curriculum Services Center in White Bear Lake, Minnesota.

**We have
a long-range interest
in good service.**

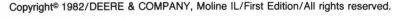
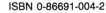

CONTENTS

Chapter 9 — SERVICING THE LUBRICATION SYSTEM

Chapter 10 — DIAGNOSIS AND TESTING OF ENGINES

PART 1

COMPONENTS AND OPERATION
OF 2- AND 4-CYCLE ENGINES

This manual is divided into two parts. Part 1 covers the components and operation of 2- and 4-cycle engines. Part 1 consists of:

Chapter 1 — Introduction To Engines

Chapter 2 — Basic 4-Cycle Engine

Chapter 3 — Basic 2-Cycle Engine

Chapter 4 — Gasoline And Diesel Fuel Systems

Part 2 consists of Chapters 5 through 10 and will cover engine service, repair, adjustment, and diagnosis.

CHAPTER 1

INTRODUCTION TO ENGINES

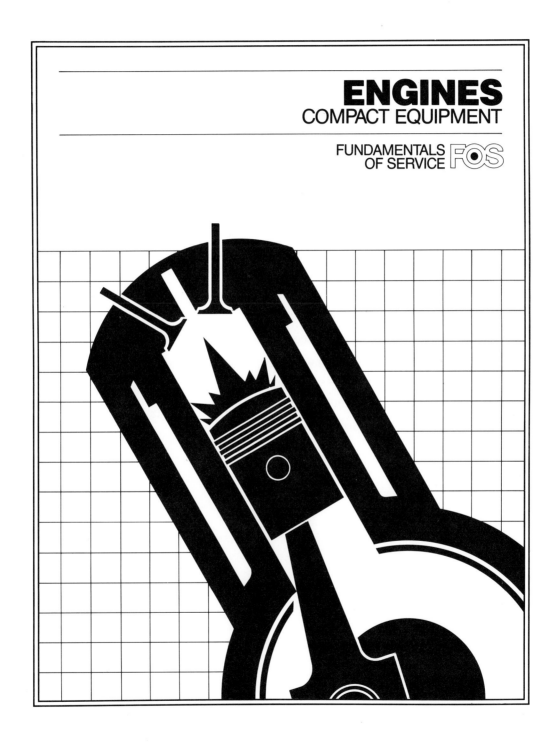

ENGINES
COMPACT EQUIPMENT

FUNDAMENTALS
OF SERVICE FOS

SKILLS AND KNOWLEDGE

This chapter contains basic information that will help you gain the necessary subject knowledge required of a service technician. With application of this knowledge and hands-on practice, you should learn the following:

• How internal combustion is the source of engine power.

• The different types of fuel delivery.

• The basic requirements of all internal-combustion piston engines.

• The four strokes of a 4-cycle engine.

• How 4-cycle engines are classified.

• How different types of lubrication systems work.

• Why there are different types of lubrication systems.

• The two strokes of a 2-cycle engine.

• The types of 2-cycle engines.

• How lubrication is accomplished in a 2-cycle engine.

• The differences between liquid-cooling systems and air-cooling systems.

• The parts of recoil and windup starters.

• How recoil and windup starters work.

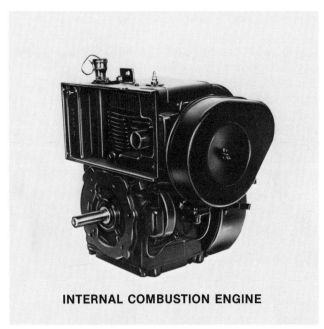

INTERNAL COMBUSTION ENGINE

Fig. 1 — Internal Combustion Piston Engine

THE INTERNAL-COMBUSTION PISTON ENGINE

A Two-cycle

An engine is a device which converts energy into mechanical motion to do work. Although there are many different types of engines, this text will deal with the internal-combustion engine under 40 horsepower (30 kW) used in compact equipment (Fig. 1).

A piston engine consists basically of a cylinder, closed at one end, and a closely fitted sliding member (piston) which is free to slide in the cylinder (Fig. 2). Pressure introduced in the closed end, forces the piston down the cylinder. If the pressure is sufficient, the force from the moving piston can be used to do useful work in addition to moving the piston.

There are many variations in the details of various types of engines but, basically, in an internal combustion engine, a mixture of air and fuel is introduced at the closed end of a cylinder and is burned within the cylinder. As combustion occurs, the gases expand and the pressure within the cylinder increases rapidly. This increased pressure acts against the piston forcing it to move down in the cylinder (Fig. 2).

In a piston engine, the reciprocating (back-and-forth or up-and-down) motion of the piston must be changed into rotating motion at an output shaft (Fig. 3). There are many ways to do this—by mechanical linkages, cams, swashplates, and eccentrics—but the most common method is by a connecting rod and crankshaft. The connecting rod is attached to an offset from the

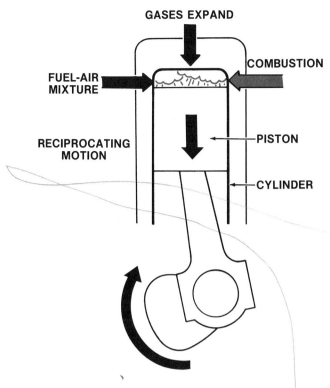

Fig. 2 — Gases Expand During Combustion

rotating centerline of the crankshaft. The result is a cranking motion similar to pedaling a bicycle. The rotary motion of the crankshaft is used for many types of work.

Fig. 3 — Reciprocating Motion Converted to Rotating Motion

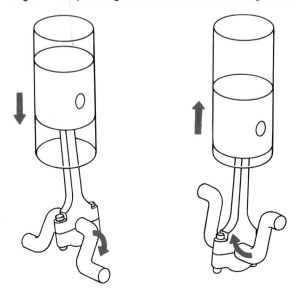

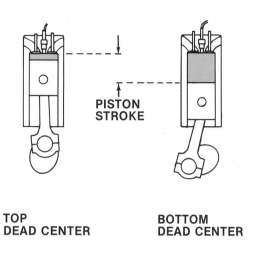

Fig. 4 — Piston Stroke

The distance of the offset is called the crank "throw" and the principle is that of a lever. Since the crankpin travels through a complete revolution, its travel to bottom center and back to top center establishes the distance the piston travels which is called the **stroke** (Fig. 4).

FUEL DELIVERY AND IGNITION

The gasoline engine uses a carburetor to mix air and fuel outside the cylinder. The mixed charge is then drawn through a tube (manifold) into the cylinder by the partial vacuum created by the intake stroke of the piston. As the piston moves back up, the charge is ignited near the top of the compression stroke by an electric spark (Fig. 5, left). This sequence will be defined in detail under the heading "How The 4-Cycle Engine Works" on the next page.

The diesel engine draws in only air on the intake stroke. The temperature of this air charge is raised to 800°F (427°C) or higher by the heat of compression. At a point near the top of the compression stroke, fuel is injected into the cylinder under high pressure. The fuel mixes with the air, and the heat caused by compression is sufficient to ignite the charge (Fig. 5, right).

A more complete discussion of fuel delivery appears in Chapter 4.

DESIGN OF ENGINE PARTS

Diesel engines must be built stronger than gasoline engines to withstand the greater forces of compression and combustion. Cylinder heads, pistons, piston pins, connecting rods, crankshafts, and bearings are all

Fig. 5 — Ignition Sequence — Gasoline and Diesel

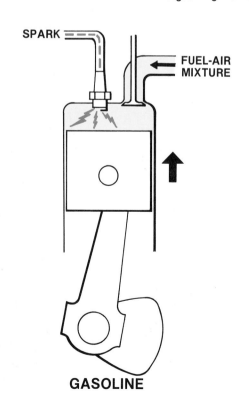

GASOLINE

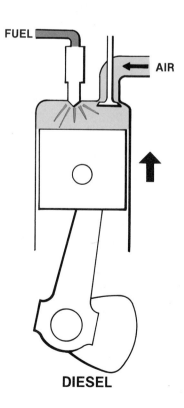

DIESEL

designed to withstand the increased loads. Design of parts also depends on whether the engine is 4-cycle or 2-cycle. Each of these types will be discussed in the following sections.

THE 4-CYCLE ENGINE

The 4-cycle engine (technically called the 4-stroke-cycle engine) illustrates the basic needs of all internal combustion piston engines. These needs are:

• *A closed combustion chamber*

• *Air to support combustion*

• *Compression of the air*

• *A means of mixing fuel with the air*

• *A means of starting combustion (ignition)*

• *Expansion of the working gases*

• *Reciprocating and rotary motion*

• *A means of exhausting the burned fuel-air residues*

• *All of the above in a timed sequence of events*

HOW THE 4-CYCLE ENGINE WORKS

The 4-cycle engine requires four strokes of the piston (two up and two down) and two revolutions of the crankshaft to complete one combustion cycle and provide one power impulse. The four strokes, and the description of their accompanying events (Fig. 6), are:

• **Intake** — The intake stroke occurs when the piston is moving down. The fuel-air mixture (air only if diesel) enters the cylinder through the open intake valve to fill the space left above the piston. The intake valve closes at about the end of the stroke.

• **Compression** — The compression stroke follows the intake stroke and occurs during the next upward movement of the piston. Intake and exhaust valves on that cylinder are closed. The fuel-air mixture (only air if diesel) is compressed to a fraction of its original volume. The mixture is ignited (fuel is injected if diesel) and combustion begins slightly before the top end of this stroke.

• **Power** — The power stroke occurs on the next downward movement of the piston. With both valves still closed, the piston is driven down by the pressure of the burning fuel-air charge. The exhaust valve opens near the bottom end of the power stroke.

• **Exhaust** — On its next upward movement the piston forces spent gases out through the open exhaust valve. The exhaust valve remains open until the piston passes the top of its travel.

COMPRESSION RATIO

The compression ratio compares the volume of a cylinder above a piston with the piston at the bottom of its stroke to the volume with the piston at the top

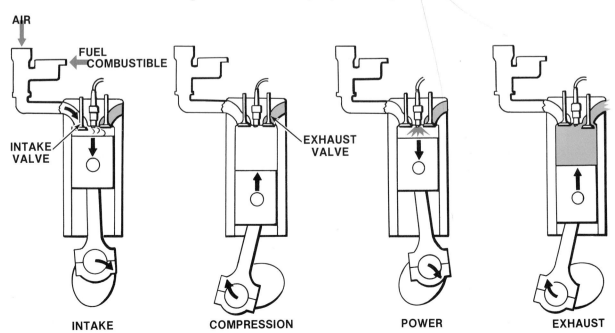

Fig. 6 — Four Strokes of Cycle (Gasoline Engine Shown)

AIR

FUEL COMBUSTIBLE

INTAKE VALVE

EXHAUST VALVE

INTAKE   COMPRESSION   POWER   EXHAUST

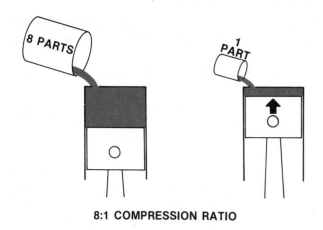

8:1 COMPRESSION RATIO

Fig. 7 — Compression Ratio

of its stroke. The comparison is expressed as a ratio (such as 8:1) which means the volume is eight times greater when the piston is at the bottom of its stroke than when it is at the top (Fig. 7).

As a gas is compressed, its pressure and temperature both increase. A high compression ratio results in greater expansion of gases, and therefore high temperature and pressure, in the cylinder during combustion which delivers a more powerful stroke. This, however, can also result in detonation or preignition. **Detonation** is an explosion of the fuel-air mixture instead of controlled burning and the resulting forces can damage an engine. **Preignition** is the premature ignition of the fuel-air charge by heat of compression which causes irregular combustion.

Consequently, an engine compressing a mixture of fuel and air, such as a typical gasoline engine, will have a

Fig. 8 — Compression Ratio — Gasoline vs. Diesel

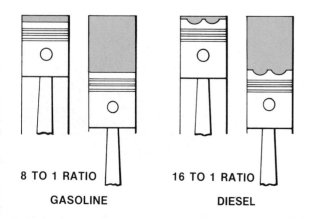

8 TO 1 RATIO

GASOLINE

16 TO 1 RATIO

DIESEL

lower compression ratio than an engine burning diesel fuel which compresses only air and injects the fuel later. Also, the diesel engine is a compression-ignition engine while gasoline engines are spark-ignition. Higher compression is necessary on the diesel engine to obtain higher combustion-chamber temperatures necessary to ignite the fuel. In practice, a typical gasoline engine may have an 8:1 compression ratio while a typical diesel will use 16:1 or higher (Fig. 8).

CLASSIFICATION OF 4-CYCLE ENGINES

All 4-cycle engines use the same sequence of strokes regardless of the mechanical arrangement of the working parts. Engines are often described or "classified" according to the arrangements. For example, engines can be classified according to:

- *Valve arrangement*
- *Cylinder arrangement*
- *Horizontal or vertical crankshaft*
- *Fuel used*

Valve Arrangement

The intake and exhaust valves may be located in the engine block or the cylinder head. Two common arrangements are called L-head and I-head (Fig. 9). The different arrangements require different mechanisms to open and close the valves. The mechanisms may be:

- *Cam and lifter in the block*
- *Cam, lifter, push rod, and rocker arm valve train*
- *Overhead cam and lifters*
- *Overhead cam, rocker arms, and lifters*

These parts will be discussed in Chapters 2 and 3.

Fig. 9 — Valve Arrangements

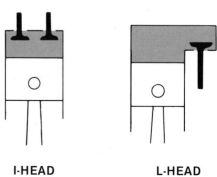

I-HEAD   L-HEAD

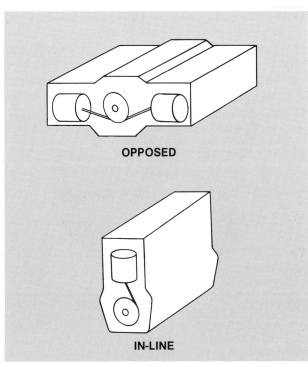

Fig. 10 — Typical Cylinder Arrangements

Cylinder Arrangement

Small engines may be single or multiple-cylinder types. Commonly, multiple-cylinder engines may be (Fig. 10):

• **In-line** — With all cylinders in a straight row above or below the crankshaft.

• **Opposed** — With two banks of cylinders on opposite sides of the crankshaft.

On multiple-cylinder engines, the cylinders must fire in a certain order as they approach top dead center on the compression stroke so the firing impulses will be evenly spaced during crankshaft rotation. There may be variations in firing order depending on the cylinder arrangement. Firing order affects engine service procedures and tune up and will be discussed in Chapter 5.

The single-cylinder arrangement is most common and is used on lawn mowers, riding mowers, and even small tractors (Fig. 11). In-line engines are found in from two- to four-cylinder arrangements on small tractors, snowmobiles, farm equipment, and standby power generators. Opposed engines are normally found on small tractors and power generators and are usually two cylinders.

Fig. 11 — Single-Cylinder and In-Line Cylinder Applications

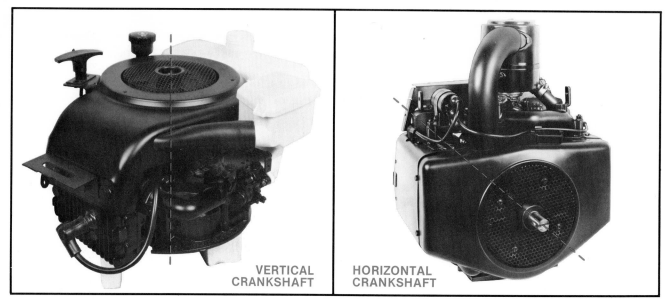

Fig. 12 — Vertical and Horizontal Crankshaft

Crankshaft Position

The crankshaft position of an engine refers to the placement of the crankshaft relative to the ground in normal operating position. A horizontal crankshaft is parallel to the ground while a vertical crankshaft is perpendicular (Fig. 12). Horizontal-crankshaft engines are usually found on garden tractors and farm machinery while the vertical-crankshaft versions are common on rotary lawn mowers and riding mowers.

Fuel Used

The type of fuel to be used plays a major part in the design and construction of an engine. Both diesel and gasoline 4-cycle engines operate on the same principles of using combustion-chamber ignition to convert energy into mechanical motion. The actual construction and operation of these engines differ on a number of points. These will be discussed in Chapter 4.

Fig. 13 — Splash-Type Lubrication System

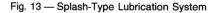
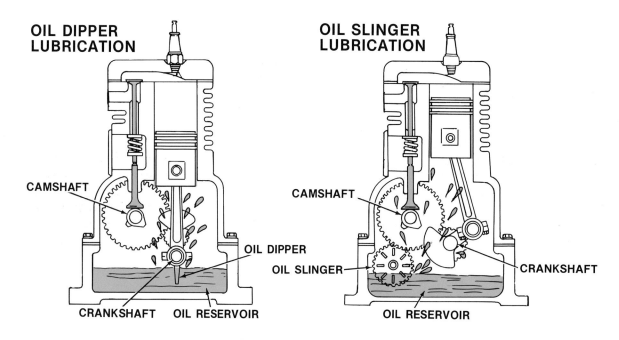

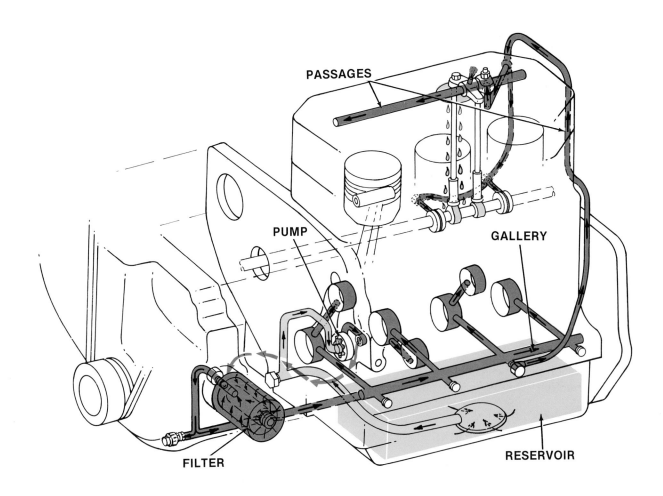

Fig. 14 — Pressure-Type Lubrication System

LUBRICATION SYSTEMS

Lubrication systems on 4-cycle engines are equipped with an oil reservoir to contain lubricating oil. The method of delivering oil to engine components depends on the type of lubrication system on the engine.

Splash-type systems (Fig. 13) use a dipper or slinger to splash oil onto engine components. More details on this system in Chapter 2. **Pressure-type** systems are more complex and contain several components.

Basic components of the pressure-type lubrication system are (Fig. 14):

- **Reservoir**
- **Pump**
- **Filter**
- **Gallery**
- **Passages**

Oil is pumped from the reservoir and contaminants are filtered out. Oil is then sent to the main oil gallery and routed through passages to lubrication points in the engine. A detailed discussion on pressure-type lubrication systems appears in Chapter 2.

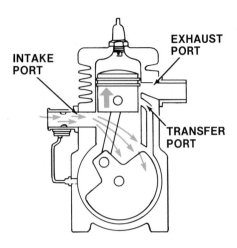

Fig. 15 — Fuel-Air Mixture Drawn Into Crankcase

THE 2-CYCLE ENGINE

Two-cycle engines share the requirements of all other internal combustion piston engines as defined earlier in this chapter under "The 4-Cycle Engine."

HOW THE 2-CYCLE ENGINE WORKS

Unlike the 4-cycle counterpart, the 2-cycle engine delivers one power impulse for each revolution of the crankshaft. Therefore, the piston goes through two strokes, instead of four, to complete one combustion cycle. To accomplish this each stroke of the piston must accomplish two functions. In 2-cycle operation:

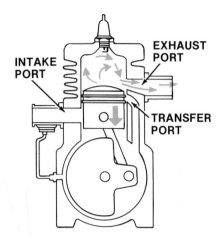

Fig. 17 — Burned Gases Escape Through Exhaust Port

• The upstroke of the piston compresses the fuel-air mixture and, by creating reduced pressure within the crankcase, draws a new fuel-air charge into the crankcase (Fig. 15).

• The downstroke of the piston is the power stroke (Fig. 16). Also, on the downstroke, burned gases, still under pressure from combustion, escape through the open exhaust port (Fig. 17). As the piston moves farther down, the transfer port is uncovered (Fig. 18). Slight pressure from the downward movement of the piston then causes the fuel-air mixture in the crankcase to move to the combustion chamber. This new charge also helps remove remaining burned gases from the combustion chamber.

Fig. 16 — Mixture Trapped in Crankcase

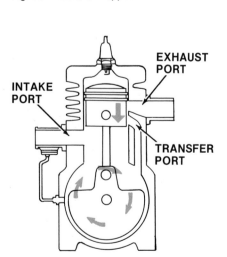

Fig. 18 — Transfer Port Uncovered

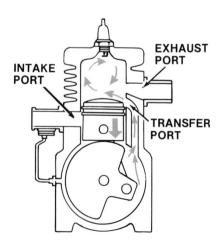

TYPES OF 2-CYCLE ENGINES

2-cycle engines for compact equipment are of three basic types:

- **Reed valve**
- **Piston ported**
- **Valve-in-head**

Reed Valve

Reed-valve engines use a reed valve in the crankcase to admit the fuel-air mixture from the carburetor (Fig. 19). When the piston is on the compression stroke, a slight vacuum is created in the crankcase which causes the reed valve to flex inward (Fig. 20) admitting the fuel-air mixture to the engine (Fig. 21). As the piston moves downward on the power stroke the pressure in the crankcase increases enough to seal the reed valve (Fig. 22). The fuel-air mixture is transferred to the combustion chamber, as described earlier, and scavenges the exhaust gases out through the exhaust ports (Fig. 23).

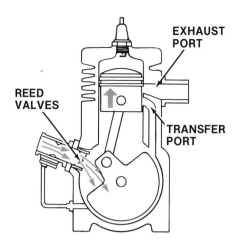

Fig. 21 — Mixture Drawn Into Crankcase

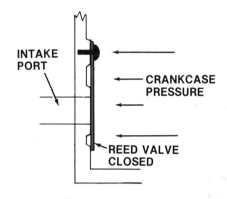

Fig. 22 — Crankcase Pressure Closes Reed Valve

Fig. 23 — Burned Gases Scavenged From Combustion Chamber

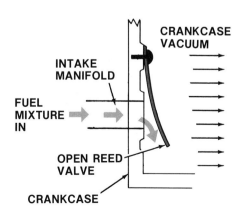

Fig. 19 — Reed Valve Components

Fig. 20 — Reed Flexes To Admit Fuel-Air Mixture

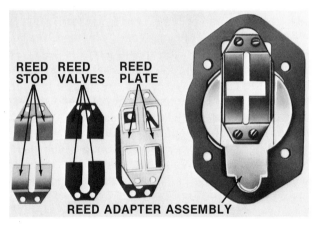

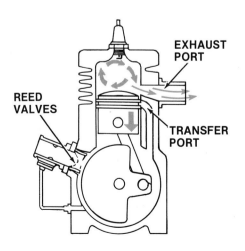

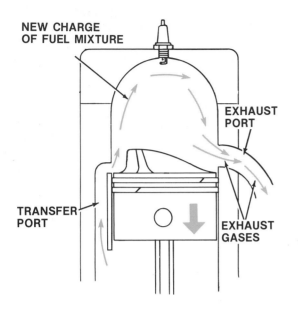

NEW CHARGE
OF FUEL MIXTURE

EXHAUST
PORT

TRANSFER
PORT

EXHAUST
GASES

Fig. 24 — Cross Scavenging

On cross-scavenged engines the piston has a ridge or deflector across the top to deflect the flow of fuel-air mix upward and create turbulence (Fig. 24). This deflected mixture forces the exhaust gases out of the pockets in the combustion chamber and down to the exhaust ports, effectively purging the cylinder of exhaust gases.

Another form of reed-valve engine uses the loop charge to scavenge the exhaust gases. In this case the

Fig. 25 — Loop Scavenging

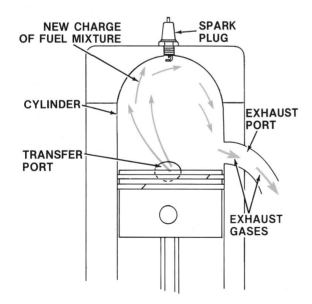

NEW CHARGE
OF FUEL MIXTURE

SPARK
PLUG

CYLINDER

EXHAUST
PORT

TRANSFER
PORT

EXHAUST
GASES

intake charge is directed upward into the combustion chamber through transfer ports which are 180 degrees apart and located on the circumference of the cylinder. The exhaust ports are also in the cylinder walls opposed and at 90 degrees to the intake ports. The exhaust ports are uncovered first on the power stroke and the exhaust gases begin to exit through the ports under their own pressure. As the piston travels farther down the cylinder the intake ports open and the fuel-air mixture, directed into the top of the combustion chamber, completes the scavenging process (Fig. 25). The top of the piston in a loop-scavenged engine does not have the ridge found on the cross-scavenged version. This allows for a lighter weight piston that is less subject to distortion. Pistons with flat crowns also run a little cooler than deflector pistons because there is 10 to 15 percent less piston surface area exposed to combustion flame.

Piston Ported

Piston-ported (or 3-port two-cycle) engine operation was described earlier in this chapter under the heading "How The 2-Cycle Engine Works." Instead of a reed valve, this type of 2-cycle engine contains a third port in the cylinder wall. As the piston skirt clears the third port on the compression stroke the fuel-air mixture is forced into the crankcase by atmospheric pressure against the reduced crankcase pressure. As the piston moves down on the power stroke the third port is closed and the descending piston pressurizes the crankcase charge. When the intake ports open the pressurized charge is forced into the combustion chamber. Like the reed-valve engine, the piston-ported engine can be either cross- or loop-scavenged.

Valve-in-Head

Valve-in-head 2-cycle engines are usually diesel engines and are usually used in heavy-duty applications beyond the scope of this book.

LUBRICATION SYSTEMS ON 2-CYCLE ENGINES

Two-cycle engines contain no oil reservoir inside the engine. Instead, gasoline and 2-cycle engine oil enter the engine together with air as a vapor. As gasoline evaporates; the oil film clings to internal engine parts. During fuel transfer, the oil moves to the combustion chamber to lubricate the cylinder wall. Two-cycle diesel engines are very uncommon for the size of engines covered in this text. A more detailed discussion of 2-cycle lubrication systems appears in Chapter 3.

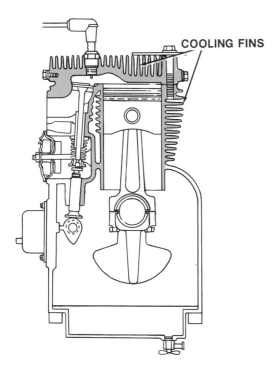

Fig. 26 — Cooling Fins on Cylinder and Head

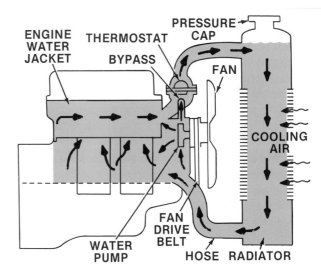

Fig. 28 — Liquid-Cooled Engine

ENGINE COOLING — 4-CYCLE AND 2-CYCLE ENGINES

An internal-combustion engine converts only part of the heat developed during combustion to mechanical power. The remainder of the heat is exhausted or absorbed by the engine components and must be carried away or engine failure will result.

Fig. 27 — Shroud Directs Air Flow

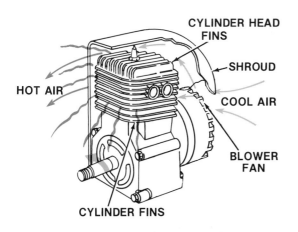

There are two basic methods to remove this heat from the engine:

- **Air cooling**
- **Liquid cooling**

Either method may apply to 2-cycle or 4-cycle engines depending on engine design.

AIR COOLING

An **air-cooled engine** is equipped with cooling fins on the cylinder head and cylinder (Fig. 26). These fins radiate heat into the surrounding air, but, in many cases, are not capable of dissipating sufficient heat. In these instances, a fan or blower is added to increase the air flow, and shrouding is provided to route the air flow where it is needed (Fig. 27). An air-cooled engine should be used where light weight and low cost are needed.

LIQUID COOLING

A **liquid-cooled engine** is equipped with a water jacket and passages around the cylinder and through the cylinder head (Fig. 28). In most cases the jacket and passages are integral parts of the engine castings. Liquid-cooled engines contain more components than air-cooled engines. Parts of a liquid cooling system are:

- **Radiator with expansion tank**
- **Water pump**
- **Radiator hoses**
- **Thermostat**

When an engine stops, the water pump also stops pumping coolant through the engine. Therefore, some coolant remains in the radiator and some remains in the engine water jacket. After the engine cools to near ambient temperature and is restarted, the thermostat will remain closed. This is to warm the engine to operating temperature as quickly as possible by not allowing the coolant to flow through the radiator.

Coolant is pumped through the water jacket passages by the water pump and picks up heat from the engine. When the coolant warms to a prescribed temperature, the thermostat opens allowing warm coolant to flow to the radiator. Heat then dissipates into the atmosphere and cooled liquid returns to the engine for cooling.

Water cooling is more efficient than air cooling but costs more and usually results in a heavier engine.

RECOIL AND WINDUP STARTERS

Recoil and windup starters are used in many gasoline engine applications. They both operate on the principle of using a manual mechanism (starter rope or return spring) to rotate the crankshaft for engine starting. Electrical starting systems are covered in a separate manual *FOS Compact Equipment Electrical Systems*.

RECOIL STARTERS

Recoil starting mechanisms usually consist of the following components (Fig. 29):

- **Starter rope**

- **Pawls, friction shoes, or eccentrics**

- **Crankshaft or flywheel adapter**

- **Return spring**

When the operator pulls the starter rope, the pawls extend out by centrifugal force to engage and turn the crankshaft or flywheel adapter (Fig. 30). The adapter then engages the crankshaft or flywheel causing it to rotate. As the crankshaft turns, the fuel and electrical systems are activated causing the engine to start.

At the end of the stroke or when the engine starts, the crankshaft adapter is disengaged by action of the pawl springs or by the action of the engine crankshaft throwing the adapter out of engagement. The rope is rewound on its pulley by the return spring.

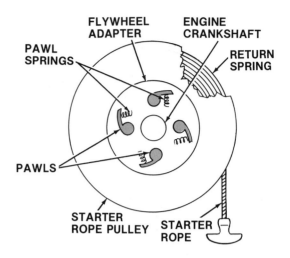

Fig. 29 — Recoil Starting Mechanism

WINDUP STARTERS

Windup starting mechanisms operate on the same basic principles as recoil starters. Instead of a starter rope, however, windup starters use a hand crank and ratchet mechanism to wind up the return spring. The return spring for windup starters is usually stronger than on recoil starters because it is used to turn the crankshaft.

Fig. 30 — Recoil Starter Operation

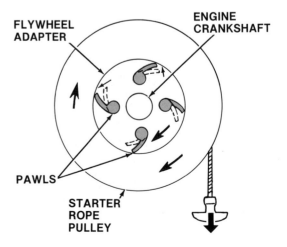

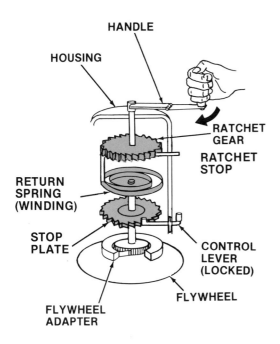

Fig. 31 — Hand Crank Winds Starter

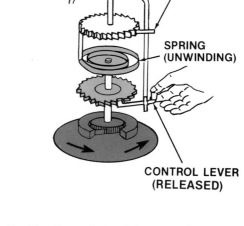

Fig. 33 — Return Spring Released for Starting

When starting an engine with a windup starter, the operator turns the hand crank to tighten the return spring (Fig. 31). This is done either directly or through a series of gears (Fig. 32). The gear-type windup makes it easier to turn the hand crank, but also requires more turns to completely wind the spring. In both types, a ratchet mechanism is used to retain the outside end of the return spring and to keep the hand crank from rebounding when the operator releases it. When the

Fig. 32 — Gear-Type Windup Starter

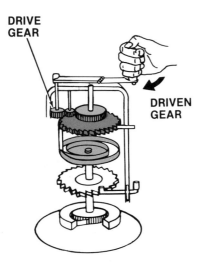

spring is tightly wound the operator actuates a control lever which releases the inside end of the spring (Fig. 33). As the return spring unwinds, it turns a shaft on which the pawls are mounted. This causes the pawls to extend and engage the flywheel adapter which then rotates the flywheel and crankshaft.

When the engine starts, the crankshaft adapter disengages and the return spring remains in a relaxed position.

Some recoil and windup starting systems are equipped with a **compression release mechanism** to make it easier for the return spring to turn the crankshaft. When the operator actuates the mechanism, a port opens in the combustion chamber. This relieves some pressure within the combustion chamber that occurs when the piston moves in the cylinder.

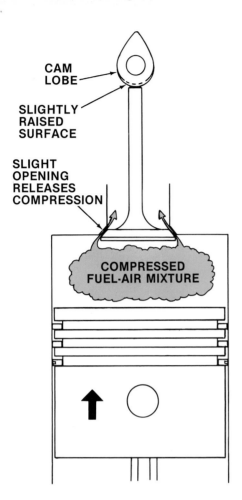

CAM LOBE

SLIGHTLY RAISED SURFACE

SLIGHT OPENING RELEASES COMPRESSION

COMPRESSED FUEL-AIR MIXTURE

Fig. 34 — Compression Release Operation

ADVANTAGES OF 4-CYCLE AND 2-CYCLE ENGINES

Both 4-cycle and 2-cycle engines are designed for certain applications. Within these applications, each has certain advantages.

4-Cycle Advantages

• *Provides better fuel economy because it operates at lower rpm*

• *Delivers more power at lower rpm*

• *Can operate in wider rpm ranges*

• *May be more durable because they do not operate at constant high speed.*

2-Cycle Advantages

• *Generally smaller and lighter with fewer moving parts*

• *Has twice as many power strokes as 4-cycle*

• *Produces more power from an equal displacement*

• *Can be mounted in various positions because the fuel mixture lubricates the engine*

Another type of compression release mechanism operates automatically. The automatic mechanism is usually found on 4-cycle engines and consists of a slightly raised surface on the camshaft lobe. This raised surface holds the intake or exhaust valve slightly open during engine cranking (Fig. 34). This system and other similar mechanisms will be covered in more detail in Chapter 2.

The ultimate result of compression release mechanisms is less resistance to the crankshaft turning. Once the engine fires or when the engine reaches a specified rpm, the compression release mechanism closes to allow normal engine compression and combustion. In some engines the compression release in a permanent orifice in the combusion chamber. After the engine begins to run, the orifice becomes ineffective in releasing pressure because of its small size.

SUMMARY

The internal combustion piston engine burns a fuel-air charge in a combustion chamber. The heat and pressure from combustion force the piston to move the crankshaft converting reciprocation motion to rotary motion.

There are two basic types of engines:

- **4-cycle**

- **2-cycle**

The 4-cycle engine requires four strokes of the piston and two revolutions of the crankshaft to complete one combustion cycle. The four strokes are **intake, compression, power, and exhaust**. Four-cycle engines are classified according to:

- **Valve arrangement**

- **Cylinder arrangement**

- **Horizontal or vertical crankshaft**

- **Fuel use**

Two types of lubrication systems available on 4-cycle engines are **splash type** and **pressure type**.

Two-cycle engines require two strokes of the piston and one revolution of the crankshaft to complete one combustion cycle. There are three types of 2-cycle engines:

- **Reed valve**

- **Piston ported**

- **Valve-in-head**

Lubrication is accomplished either by a fuel-oil premix from the fuel tank or by injecting 2-cycle engine oil into the gasoline at the air intake port.

Engine cooling is accomplished by passing ambient air over the engine or by pumping coolant through a water jacket in the engine. Air-cooled engines are usually lighter weight and less expensive but liquid-cooled engines are more efficient.

Recoil and windup mechanisms are used to start some engines. These eliminate the need for electric starting systems. In either of these systems, the operator mechanically causes the crankshaft to turn by pulling a starter rope or winding a return spring.

CHAPTER 1 REVIEW

1. (Fill in the blanks) The internal combustion engine burns _____and _____ within a _____to convert heat energy to mechanical power.

2. (Fill in the blanks) In a piston engine the _____ (back-and-forth or up-and-down) motion of the _____must be changed into _____motion at the output shaft.

3. Air is compressed in the engine cylinder. What else happens to this air?

4. Name six of the nine basic needs of all internal-combustion piston engines.

5. (Fill in the blanks) The 4-cycle engine requires _____strokes of the _____ and _____revolutions of the _____ to provide _____ power impulses.

6. Name the four strokes of the 4-cycle engine combustion cycle in the proper sequence.

7. Name the four ways 4-cycle engines can be classified.

8. Why is the compression ratio higher on diesel engines than gasoline engines?

9. What are the two major types of engine cooling systems?

CHAPTER 2

BASIC 4-CYCLE ENGINE

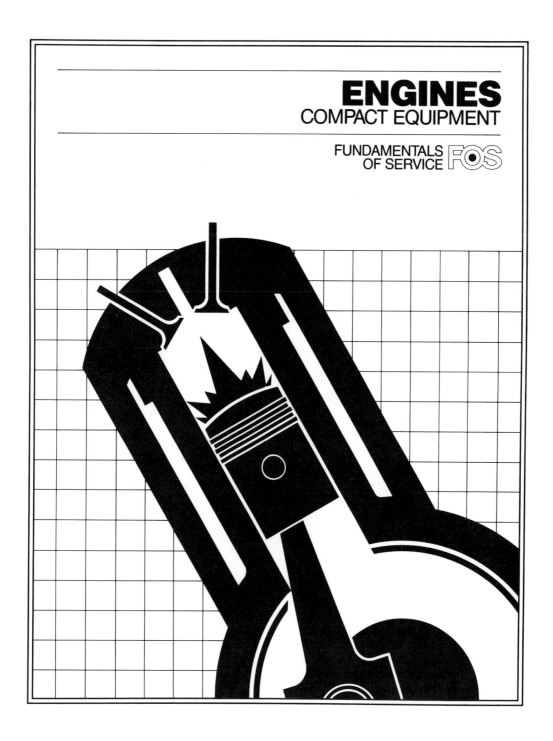

ENGINES
COMPACT EQUIPMENT

FUNDAMENTALS
OF SERVICE FOS

SKILLS AND KNOWLEDGE

This chapter contains basic information that will help you gain the necessary subject knowledge required of a service technician. With application of this knowledge and hands-on practice, you should learn the following:

• The components of a 4-cycle engine.

• The function of each component of a 4-cycle engine.

• How the components work.

• The differences between types of lubrication systems.

• The different types of oil pumps used in pressure lubrication systems.

• What types and viscosities of engine oils are required for 4-cycle engines.

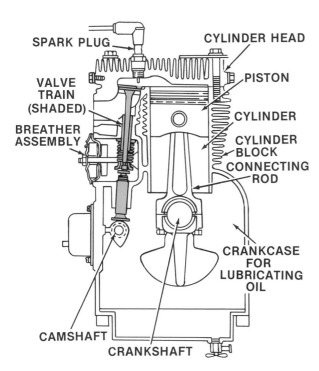

Fig. 1 — Four-Cycle Engine Components

INTRODUCTION

Engine designs can differ greatly, but the components common to all 4-cycle engines (Fig. 1) are:

- **Cylinder block**
- **Cylinder**
- **Cylinder head**
- **Piston**
- **Connecting rod**
- **Crankshaft**
- **Camshaft**
- **Valve train**
- **Lubrication system**

Each component has a specific function which will be discussed throughout this chapter. Service and repair procedures for these components will be discussed in Chapter 5.

CYLINDER BLOCK

The cylinder block (Fig. 2) is the foundation that supports the rest of the engine and holds the other components in proper relationship. Many types of cylinder blocks are available depending on the design of the engine.

Blocks on lightweight, single-cylinder gasoline engines are often cast in two pieces—a main block casting and a removable end-wall section (Fig. 2, left). During machining, the pieces are aligned with locating pins, bolted together, and the camshaft and crankshaft bearing support surfaces line-bored for accuracy. This type of construction is common in engines up to about 20 horsepower (15 kilowatts).

Blocks on multiple-cylinder engines (both gasoline and diesel) are commonly one piece (Fig. 2, right). Diesel-engine blocks are also constructed stronger because of greater forces of compression and combustion.

Cylinder blocks for both gasoline and diesel engines are usually cast of gray iron or aluminum and precision machined. They are designed to include:

- *End walls and center webs (Fig. 2) which support the crankshaft and camshaft, seal seats, and, sometimes, bearing retainers*

Fig. 2 — Cylinder Blocks — Single-Cylinder Gasoline (Left), Multiple-Cylinder Diesel (Right)

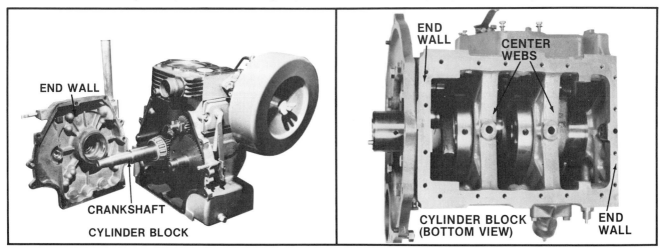

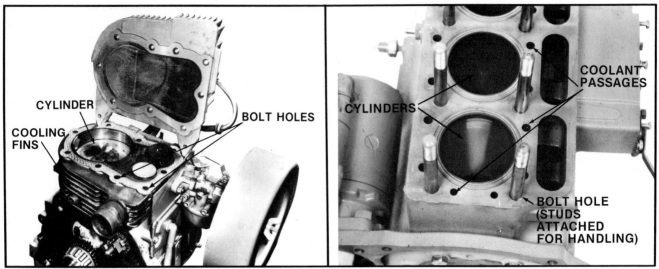

Fig. 3 — Top Cylinder Block — Single Cylinder Gasoline (Left), Multiple-Cylinder Diesel (Right)

• Cylinders (Fig. 3), cylinder counterbores (wet-sleeve engines), or cylinder-mount counterbores and reinforced stud holes (if a separate cylinder is used)

• *Running surfaces and clearance holes for valve train components and, in many cases, valve guides or guide bores, valve seats, and manifold ports*

• *Stud or bolt holes (Fig. 3) and mating surfaces for the cylinder head*

• *Coolant passages for liquid-cooled engines, or cooling fins for air-cooled engines (Fig. 3)*

• *Mounting and, in many cases, running surfaces for other engine systems*

• *Oil passages and return lines in engines with pressure-type lubrication systems*

• *Breather passages for air displaced by piston movement and blowby*

CYLINDER

The cylinder is a tube in which the piston works and compression and power are developed. The two basic types of cylinders are (Fig. 4):

• **Cast-in-block (Enbloc)**

• **Individual castings**

ENBLOC CYLINDERS

Enbloc cylinders are cast into the engine block. In aluminum-block engines, cylinder liners are sometimes cast in place or the cast bore may be machined with a steel or cast-iron cylinder liner shrunk or pressed in place. This liner provides the actual running surface for the piston. A few of the very light aluminum-block engines actually use a machined and treated aluminum bore for the running surface.

Cast-iron enbloc-type blocks normally are not equipped with cylinder liners since the iron cylinder bore has good wear qualities. Exceptions, however, are cast-iron blocks with thin-wall liners of cast iron to simplify rebuilding.

INDIVIDUALLY CAST CYLINDERS

Engines equipped with individually cast cylinders are usually wet-sleeve-in-block.

Fig. 4 — Types of Cylinder Liners

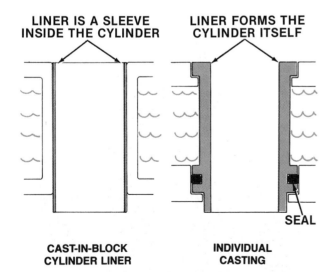

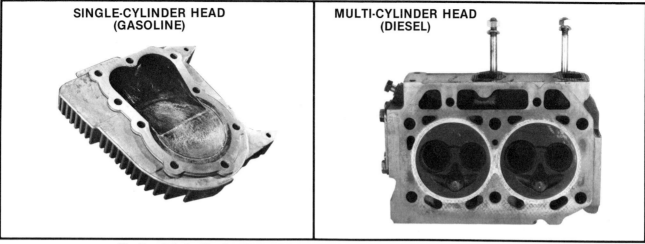

SINGLE-CYLINDER HEAD
(GASOLINE)

MULTI-CYLINDER HEAD
(DIESEL)

Fig. 5 — Cylinder Head — Single-Cylinder Gasoline (Left), Multiple-Cylinder Diesel (Right)

A wet-sleeve-in-block engine uses a separately cast and machined iron cylinder inserted into a counterbore in the cylinder block. Since the outside of this sleeve forms the inside of the engine water jacket (hence the name wet sleeve), the sleeve is usually equipped with flanges and some form of gasket or seal ring at each end to prevent coolant leaks. The inside of the sleeve is a conventional cylinder which may be removed and replaced if damaged or worn excessively.

CYLINDER HEAD

The primary purpose of the cylinder head is to seal one end of the cylinder so that the piston can develop compression and power. Again, because of engine requirements for spark ignition or fuel injection, and because of single- and multiple-cylinder designs, cylinder head appearance varies (Fig. 5). The contour of the cylinder head must be carefully matched to the requirements of the engine combustion chamber. Thus, cylinder heads of many patterns exist, but they all share some common features:

• *A means of sealing tightly against the cylinder*

• *A combustion chamber*

• *Accommodation for spark ignition or fuel injection*

• *Provision for cooling (and lubrication in overhead valve applications)*

CYLINDER-HEAD SEALING

The cylinder head must be tightly sealed against the cylinder to prevent leaking and a resulting loss of compression and power. This seal is usually accomplished by placing a gasket between mating flat surfaces on the cylinder and cylinder head (Fig. 6).

When the head is bolted tightly to the cylinder, the gasket is crushed between the two surfaces and forms a tight seal. This prevents the compressed gases from escaping and causes them to exert their maximum force on the piston.

COMBUSTION CHAMBER

The combustion chamber is the confined volume between the cylinder head and the top of the piston. The cylinder head portion of the combustion chamber is a concave or flat area on the bottom of the head. It is shaped to:

• *Allow the charge from the intake valve to reach the cylinder*

Fig. 6 — Gasket Between Cylinder Head and Block

HEAD
GASKET

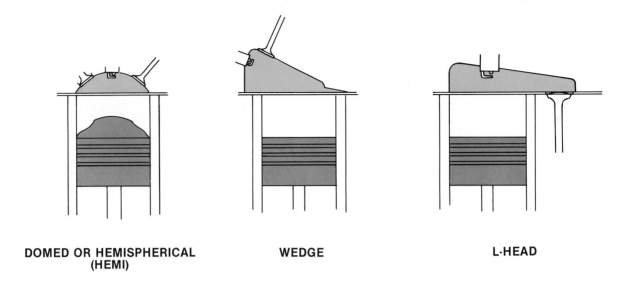

DOMED OR HEMISPHERICAL
(HEMI)   WEDGE   L-HEAD

Fig. 7 — Valve Layouts

• *Allow the spent gases to reach the exhaust valve*

• *Provide the necessary controlled fuel and air flow and turbulence for proper combustion*

• *Provide operating clearance for the valves*

Obviously, different valve layouts and engine designs will affect the shape of the combustion chamber and the size (total volume) will be determined by the desired compression ratio (Fig. 7).

Some cylinder heads (notably those designed for diesel use) may have an additional chamber for precombustion of fuel cast into or inserted in the head adjacent to the main combustion chamber (Fig. 8).

The purpose of such a chamber is to aid in the efficient and complete combustion of the fuel charge.

COMBUSTION CHAMBER IGNITION

Cylinder heads for gasoline engines normally have a threaded hole from the outside of the head to the inside of the combustion chamber (Fig. 9). This hole is designed to accommodate a spark plug which is used to ignite the fuel-air mix electrically.

Heads designed for diesel applications may have a similar hole for a fuel injection nozzle. The nozzle sprays fuel directly into the combustion chamber or may be designed so the injector sprays the fuel into the precombustion chamber as shown in Fig. 9, right.

COOLING AND LUBRICATION

Air-cooled cylinder heads incorporate cooling fins to transfer heat to the air (Fig. 10). If the engine uses a blower to move the air and shrouding to direct it, the cylinder-head fins are arranged to take advantage of the available air flow.

Liquid-cooled cylinder heads contain internally cast passages to match the passages in the cylinder block. This allows a free flow of coolant through the block and head. The head gasket must, of course, have matching passages or coolant flow would be cut off (Fig. 11). This makes it critical that the head and head gasket are installed correctly to avoid blocking these passages.

Valve-in-head engines require lubricating oil for cylinder head components. Lubrication systems will be discussed in more detail later in this chapter.

Fig. 8 — Precombustion Chambers

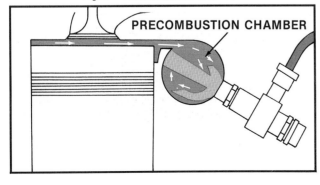

PRECOMBUSTION CHAMBER

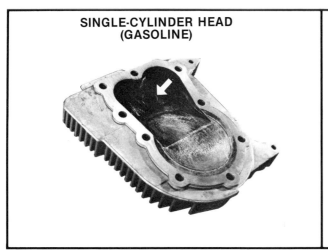

SINGLE-CYLINDER HEAD (GASOLINE)

MULTI-CYLINDER HEAD (DIESEL)

Fig. 9 — Gasoline Engine Cylinder Head (Left), Diesel Cylinder Head (Right)

COOLING FINS

Fig. 10 — Air-Cooled Cylinder Head

Fig. 11 — Coolant Passages in Head Gasket

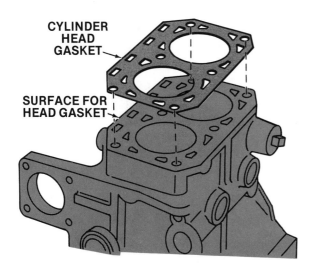

CYLINDER HEAD GASKET

SURFACE FOR HEAD GASKET

PISTON

The piston slides in the cylinder and is fitted closely to the cylinder to seal the compression. Since the piston and cylinder are heated and may expand at different rates, there must be a flexible member—the piston rings—to accept the differences.

PISTON RINGS

Piston rings are cast iron, or combinations of cast iron and steel. They fit into grooves machined into the circumference of the piston (Fig. 12). When the piston and rings are installed in the cylinder, the rings perform three functions:

- *Expand against the cylinder walls to form a gas-tight seal between piston and cylinder*

- *Help cool the piston by transferring heat*

- *Control lubrication between piston and cylinder wall*

Fig. 12 — Piston Ring Grooves

RING GROOVES
COMPRESSION RINGS

RING GROOVE
OIL CONTROL RING

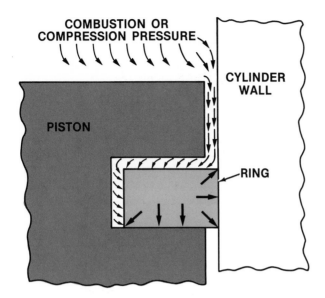

Fig. 13 — Combustion Helps Rings Seal

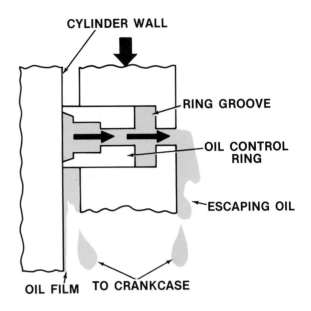

Fig. 15 — Oil Control Ring Operation

The clearance between rings and piston grooves allows the piston to expand without scuffing the cylinder wall.

The two basic types of piston rings are:

- **Compression**

- **Oil control**

Fig. 14 — Types of Compression Rings

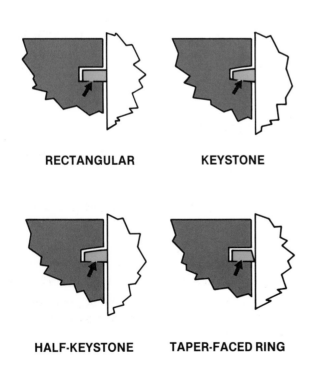

RECTANGULAR   KEYSTONE

HALF-KEYSTONE   TAPER-FACED RING

Compression rings seal the cylinder for maximum compression. They are the ring or rings nearest the top of the piston and expand out against the cylinder wall to provide the seal. Combustion pressure also helps to assure a good seal by forcing the ring against the bottom of the ring groove and out against the cylinder wall (Fig. 13). Compression rings remove small carbon deposits from the cylinder walls during operation.

Four types of compression rings are used on 4-cycle engines. The type used depends on engine design.

Rectangular rings contact the cylinder wall along the entire face (Fig. 14). **Keystone and half-keystone rings** (Fig. 14) help to reduce deposits that may fill ring

Fig. 16 — Parts of Piston Body

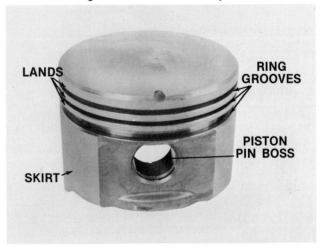

grooves and cause sticking. They also contact the cylinder wall along the entire face. **Taper-faced rings** (Fig. 14) contact the cylinder wall with only the lower outside edge. They seat quickly and provide good oil wiping on the downstroke.

The **oil control ring** is the bottom ring on a piston and wipes the excess oil from the cylinder wall. This oil is routed through slots or holes in the oil ring to holes in the oil ring groove of the piston which feed the oil back to the crankcase (Fig. 15).

PARTS OF THE PISTON BODY

The parts of the piston body are (Fig. 16):

• **Crown** — The top surface which receives the pressure of combustion. The crown may be shaped to increase compression, swirl the charge in the manner of a turbulence chamber, and, in some cases, provide clearance for the valves. The shape of the crown also depends on whether the engine is gasoline or diesel.

• **Ring grooves** — Carry the piston rings

• **Lands** — The raised areas between the ring grooves which aid in containing and applying pressure to the rings. They receive the force of the combustion gases as applied to the rings.

• **Piston pin boss** — Accepts the piston pin and provides the force-bearing surfaces used to transmit power and motion to the connecting rod.

• **Piston pin** — Connects the piston to the connecting rod (Fig. 17). The pin may rotate freely in both connecting rod ends and pin bosses (full-floating pin). Piston pins require retainers to keep them in position within the bosses. Another type of pin may be press fit in connecting rod eye.

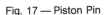

Fig. 17 — Piston Pin

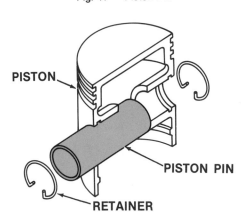

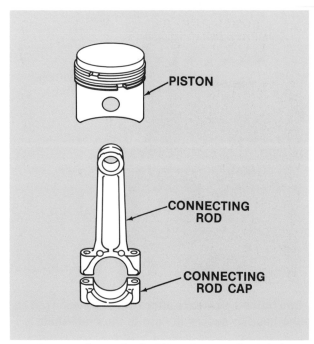

Fig. 18 — Connecting Rod Links Piston to Crankshaft

• **Piston skirt** — The cylindrical area below the ring grooves which may be ground to a slightly elliptical and tapered shape. This shape allows the piston to expand to a cylindrical shape as it heats up. This results in a better gas seal and better fit for less contact between the skirt and the cylinder walls. Pistons used in the lighter and less expensive engine applications may be completely cylindrical instead of cam ground.

The piston must be properly weighted and balanced to provide smooth engine performance. This is especially true with multiple cylinder configurations where the cylinders must balance each other precisely to avoid excessive engine vibration.

CONNECTING ROD

The connecting rod links the piston to the crankshaft (Fig. 18). The parts of the connecting rod (Fig. 19) and their function are:

• **Eye** — Accepts the piston pin and provides the direct link to the piston.

• **Head and cap** — The direct connection to the crankshaft. They are cut from a single piece head and fitted with the crankpin bearings if bearings are used.

• **Shank** — The solid, normally tapered, link between head and eye which is cast or forged as a part of the connecting rod.

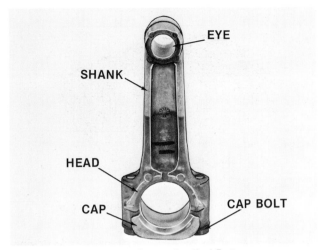

Fig. 19 — Connecting Rod Parts

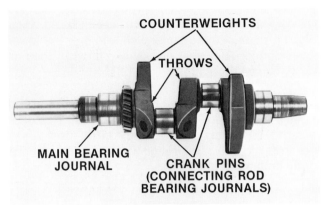

Fig. 20 — Crankshaft Parts

- **Cap bolts** — Used to attach the connecting rod cap to the head to secure the rod to the crankshaft.

- **Lock tabs** — A machined part of the cap which secures cap bolts.

The connecting rod eye is equipped with a bushing or bearing closely fitted to the piston pin. This bushing provides a thrust and wear surface for the pin.

The head and end cap are fitted with a split bearing insert set which provides the thrust and wear surfaces against the crankpin. Many engines with lightweight metal connecting rods do not use separate bearings.

CRANKSHAFT

The crankshaft (Fig. 20) converts the linear motion of the piston, through the connecting rod, to rotary motion. Since the crankshaft is subject to severe stresses it is usually a high-strength casting or forging, precision machined, balanced, and carefully inspected. The parts of the crankshaft are:

- **Journals** — Bearing surfaces used to support the crankshaft.

- **Crankpins** — Bearing surfaces which accept the thrust of the connecting rod.

- **Throws** — The offsets which support the crankpins and provide the leverage to rotate the crankshaft.

- **Counterweights** — Balancing weights opposite the crankpins.

The main-bearing journals are mounted in bearing surfaces in the front and rear walls and support webs of the cylinder block casting. The crankshaft rotates on these fixed bearing points under the thrust of the connecting rod against the crankpin.

The bearings used on the crankpins differ from the bushing used on the eye end of the connecting rod. While the bushing is a one piece cylinder the crankpin bearings are generally two piece inserts (Fig. 21). The inserts are fitted to the connecting rod head and cap and bolted around the crankpin. Lubrication holes may be present in bearing inserts. They are matched to lubrication passages in the connecting rod and cap and must be properly assembled to assure the passages are not blocked. Locking tabs are also used to prevent bearing rotation.

Antifriction bearings are often found in compact equipment engines. Such ball or roller, bearings (Fig. 22) installed on main bearings and journals, are used when minimum friction is desired. Roller bearings are polished steel cylinders arranged around a crankpin or bearing journal, usually in preformed races which contain the bearings. When the engine is running the motion of the bearing is, thus, rolling instead of sliding.

Single-cylinder engines of the type used on lawn mowers and some compact utility tractors are usually arranged with bushings (or ball or tapered-roller bearings) in the end walls of the cylinder block which act

Fig. 21 — Plain Bearings or Inserts

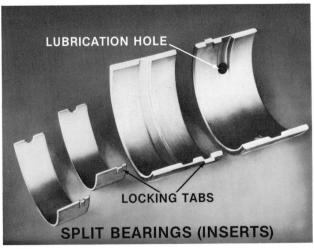

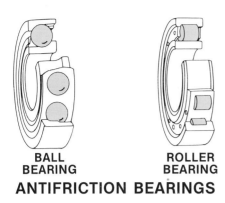

BALL BEARING   **ROLLER BEARING**

ANTIFRICTION BEARINGS

Fig. 22 — Antifriction Bearings

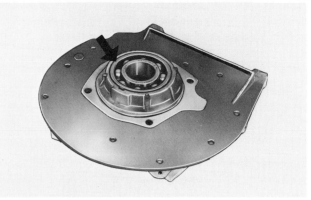

Fig. 23 — Ball Bearings Act as Main Bearings

as main bearings (Fig. 23). (The two-piece block is assembled around the crankshaft.)

The crankshaft throw positions affect:

- **Engine balance**
- **Torsional vibration of the crankshaft**
- **Main-bearing loading**
- **Engine firing order**

Therefore, the crankshaft throws are statically and dynamically balanced to keep vibration at a minimum and main-bearing loading as constant as possible.

FLYWHEEL

The flywheel is a disk of iron, steel, or aluminum bolted and sometimes keyed to one end of the crankshaft.

The design of the flywheel depends on engine design and whether the engine is air- or liquid-cooled (Fig. 24). It is used to:

- *Store energy between power pulses of the enigne*
- *Stabilize the speed of the crankshaft*
- *Transmit power to the load*
- *Act as a fan for some air-cooled engines*

The flywheel provides the inertia to spin the engine through its nonpower strokes (exhaust, intake, and compression) without slowing down. In fact, a heavy enough flywheel will actually dampen the vibration of a single-cylinder engine so that there is no sign of the individual power pulses. The inertia is sufficient to carry the engine and load through 1½ revolutions between power strokes without noticeable slowing.

But, a heavy flywheel also has disadvantages. The amount of energy required to bring it to speed in-

Fig. 24 — Air-Cooled Engine Flywheel (Left), Liquid-Cooled Engine Flywheel (Right)

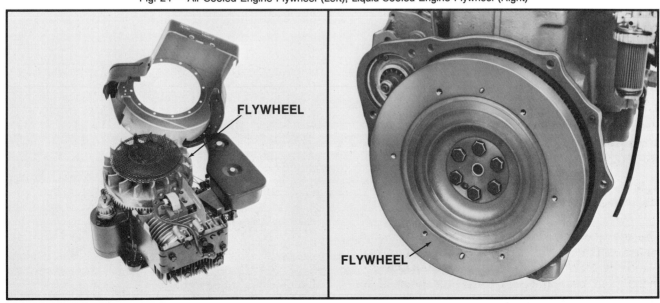

FLYWHEEL

FLYWHEEL

creases with the weight of the flywheel. Also, a heavy flywheel hampers the ability of the engine to change speed quickly.

Other common functions of the flywheel are:

• **Starting device (electric-start engines)** — A gear on the starter motor meshes with a ring gear on the flywheel and spins the flywheel (Fig. 24, right) and crankshaft until the engine starts.

• **Power line** — When fitted with a clutch and pressure plate, the flywheel is used as a direct link in the driveline.

• **Engine cooling** — Most air-cooled engines of the type used on outdoor power equipment use fins on the flywheel to act as a blower for cooling air when the engine is running (Fig. 24, left).

• **Electrical** — The magneto ignitions on many small engines use one or more magnets cast into the flywheel as a part of the ignition current and battery charging generating system.

• **Governor** — Air flow created by flywheel actuates pneumatic-type governor.

The size and weight of the flywheel used on a given engine depends on the amount of vibration and pulse damping needed. Other considerations are the number of cylinders and compression ratio of the engine, the number and types of extra systems for which the flywheel drives or supplies power, the speed response required, and the size and type of load the engine is expected to drive.

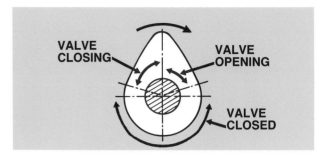

Fig. 26 — Cam Lobe Angles

CAMSHAFT

The camshaft is responsible for the timing, lift, and duration of the valve movements. It may also be used to run engine parts such as oil pump, fuel pump, and oil slingers.

A camshaft on a 4-cycle engine operates at one-half the speed of the crankshaft. It is coupled to the crankshaft by a drive system, such as gears, lugged belts and pullies, or chains and sprockets. The drive system assures that the camshaft will always be at the same point of rotation relative to the crankshaft at any given point in the engine cycle.

The camshaft itself is a round bar usually made of hardened iron. Raised polished surfaces are located

Fig. 27 — Automatic Compression Release Mechanisms

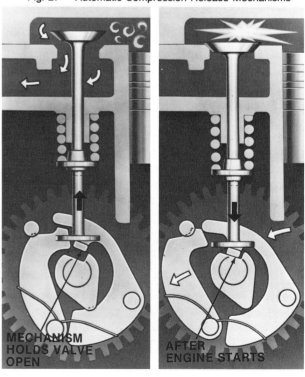

Fig. 25 — Camshaft Parts

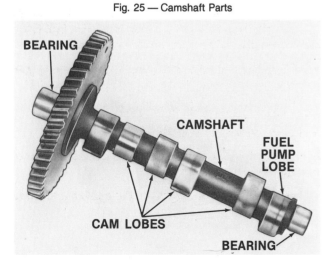

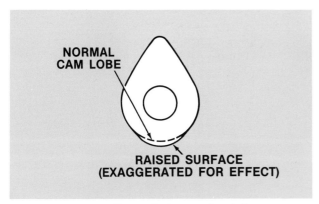

Fig. 28 — Raised Surface on Cam Lobe

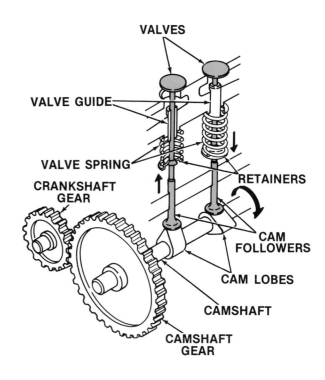

Fig. 29 — Common Valve Train Components

at intervals along the bar and are known as lobes (Fig. 25). The lobes are positioned to align with and drive some part of the valve train. The high points of the lobes are located at such an angle to each other that, as the engine turns, the valves for each cylinder are forced open and allowed to close again in the proper sequence and time. The rate at which the valve opens is determined by the angle of the lobe (Fig. 26). The distance the valve opens is controlled by the height of the lobe at its maximum. The number of degrees of camshaft rotation the lobe holds pressure on the valve determines the duration of the valve opening.

On some engines, the camshaft contains an automatic compression release mechanism. This mechanism helps reduce cranking effort by holding the exhaust valve open slightly during the first part of the compression stroke (Fig. 27, left). The crankshaft turns with less resistance because pressure is reduced in the combustion chamber.

When the engine reaches a specified rpm, centrifugal force throws the compression release mechanism out of operation (Fig. 27, right). This allows the engine to run normally at higher speeds with no power loss.

Another type of compression release mechanism consists of a slightly raised surface on the cam lobe (Fig. 28). This also holds the exhaust valve open to reduce cranking effort. This mechanism operates at all times since it is built into the cam lobe. However, at high engine rpm, the small amount that the exhaust valve is open during the compression stroke has virtually no effect on engine operation.

The camshaft not only determines the basic timing to allow the engine to operate but, by the size and shape of the lobes, determines the speed and power range at which the engine will operate most efficiently.

Depending on engine design, the camshaft may also actuate breather points (or distributor) for ignition. It may also be used to actuate the fuel pump or oil pump or to operate the governor.

VALVE TRAINS

The valve train provides the means for admitting air (and fuel in gasoline engines) to the cylinder. When the valves close the combustion chamber is sealed for compression and power. The valve train also provides a means of releasing burned gases after combustion. A valve train consists of many components and not all of them are used on a given valve train (Fig. 29). Each 4-cycle-engine valve train, however, will have:

• *At least one intake and one exhaust valve per cylinder*

• *A means of actuating the valves open*

• *A cam follower to bear against the cam lobe*

• *Valve springs to force the valves closed when not held by the cam follower*

• *A valve seat for the valves to seal against and close off the cylinder*

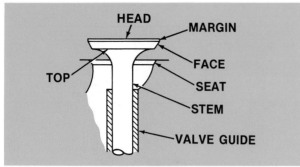

Fig. 30 — Parts of Valve

VALVES

Engine valves have two basic parts: the head and the stem (Fig. 30). The **head** of the valve is comprised of:

• **Top** — The surface which faces the inside of the combustion chamber.

• **Face** — The tapered face is the part which seals against the valve seat.

• **Margin** — The area between the edge of the top and the upper edge of the face.

The **valve stem** is the long portion of the valve which is attached to the bottom of the head. The sides of the stem are the bearing surface that runs in the **valve guide**. These surfaces keep the valve running accurately in a straight line and transfer heat away from the valve head and through the valve guide for cooling.

The outer end of the valve stem is equipped with grooves or a hole to accept **a retainer**. Retainers are devices which lock against the valve spring cap and hold the valve in position.

Fig. 31 — Types of Retainers

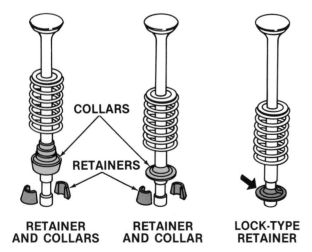

COLLARS

RETAINERS

RETAINER AND COLLARS   RETAINER AND COLLAR   LOCK-TYPE RETAINER

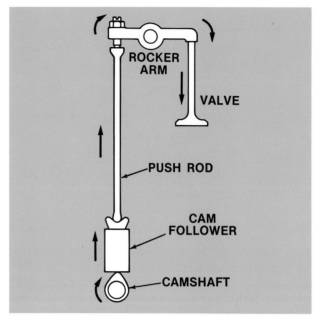

Fig. 32 — Push Rod and Rocker Arm Operation

Two types of retainers are available (Fig. 31):

• **Retainer and collar**

• **Lock type**

The retainer and collar type uses a retainer to support the spring and a set of tapered collars which slip under the retainer to hold it on the valve stem. The lock-type retainer is inserted over the valve stem and held in place by spring pressure.

The tip of the valve stem receives the force from the cam follower and camshaft to open the valve. In some cases a **rocker arm** is placed between the follower and valve and actuated by a **push rod** (Fig. 32). Rocker arm and push rod operation will be discussed under the next section ''CAM FOLLOWERS.''

Valves may be fitted with devices to rotate them and scour the combustion deposits from the seating surfaces. These devices are called **rotators**.

CAM FOLLOWERS

The cam follower, used to follow the profile of the cam lobe and actuate the valve, may be one of several different types (Fig. 33):

• **Lifter** — A cylindrical piece or assembly which slides in a bore in an engine block above the cam lobe. Lifters may be either solid or adjustable. Hydraulic self-adjusting lifters are rare on small engines.

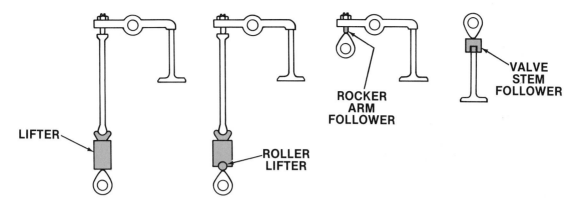

Fig. 33 — Types of Cam Followers

• **Roller lifter** — Similar to the regular lifter, but contains a roller on its cam contact surface to minimize wear. This type is rarely used.

• **Rocker-arm follower** — One end of the rocker-arm rides on the cam lobe surface and transmits the motion directly to the valve.

• **Valve-stem follower** — The valve-stem tip, or a small wear cup fitted over it, rides directly on the cam lobe surface and valve motion is accomplished by the cam. This arrangement is popular in L-head engines.

Lifter- and roller-lifter type followers usually require use of a push rod and rocker arm assembly for valve actuation (Fig. 33). The **push rod**, a straight metal rod or tube, rides on the follower at one end and seats against the rocker arm at the other end. The rocker arm, as its name implies, is an arm which pivots at the center. As the push rod forces one end of the rocker arm up, under pressure of the cam follower, the other end of the rocker arm forces the valve down and open. The push rod and rocker arm assembly are typically used in valve-in-head engines. Most compact equipment engines use lifter-to-valve actuation rather than push rods.

VALVE SPRINGS

Valve springs are cylindrical springs used to close the valves and assure proper seating. One or more springs are used per valve and the spring rate, or stiffness, is carefully selected for the application (Fig. 34).

Fig. 34 — Valve Springs

Fig. 35 — Valve Face and Seat Angles

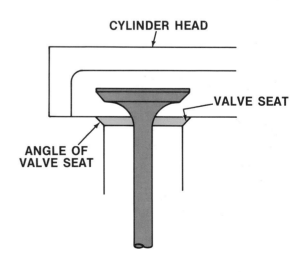

VALVE SEATS

Valve seats are the circular seats in the cylinder block or head that the face of the valve seals against. The angle of the seat matches the angle of the valve face for proper fit (Fig. 35). (Exact angles are determined by the engine manufacturer.) Valve seats may be ground directly into the head or block material, if cast iron, or may be special alloy steel inserts in either cast iron or aluminum engines.

LUBRICATION SYSTEMS

4-cycle engines on compact equipment normally use one of two types of lubrication systems:

• **Splash type**

• **Pressure type**

Engine design determines the type of system used. In any system, check oil levels daily to assure proper engine operation. Also, change oil as recommended in the operator's manual, especially for splash-type lubrication systems which have no oil filter.

SPLASH TYPE

The **oil dipper** is the simplest splash-type lubrication system. It consists of a dipper (or vane) either molded into the connecting rod cap or attached as part of the cap bolt locks. The dipper dips into the oil reservoir as the crankshaft turns throwing (or splashing) oil and oil

Fig. 36 — Oil Dipper Lubrication

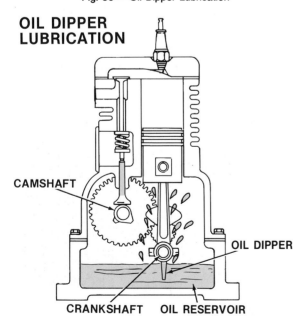

OIL DIPPER LUBRICATION

CAMSHAFT

OIL DIPPER

CRANKSHAFT   OIL RESERVOIR

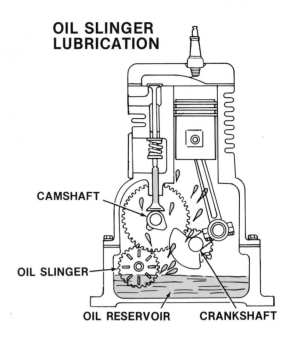

OIL SLINGER LUBRICATION

CAMSHAFT

OIL SLINGER

OIL RESERVOIR   CRANKSHAFT

Fig. 37 — Oil Slinger Lubrication

vapor into the upper part of the crankcase or block (Fig. 36). The oil lubricates and cools moving engine parts and then returns to the crankcase.

An **oil slinger** is a disk or wheel fitted with cups or plates for scooping oil. This disk is partially submerged in the oil reservoir and turned by the camshaft or crankshaft gear (Fig. 37). The plates lift oil from the reservoir and sling it into the upper part of the engine to cool and lubricate engine parts.

PRESSURE LUBRICATION

Pressure lubrication systems consist of a positive displacement oil pump which picks up oil from the oil reservoir and pumps it through internal oil galleries and a filter to engine components (Fig. 38). Most systems of this type provide oil to the crankpins (connecting-rod bearings) through passages in the crankshaft and bearings, and then route the oil through the connecting rod to the piston pin. Other systems spray oil to the piston pin rather than route oil through the connecting rod. The cam bearings are normally lubricated through passages to the bearings. Some engines may be combination positive pressure and dipper or slinger.

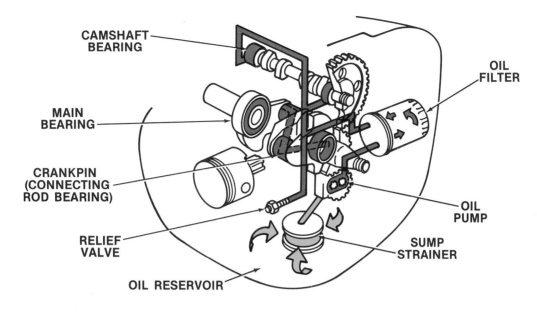

Fig. 38 — Pressure Lubrication System

Oil Pumps

There are three types of pressure oil pumps:

- **Plunger pump**
- **External gear pump**
- **Rotor pump**

All three types of pumps draw oil from the sump and force it, under pressure, into the oil passages. The outlets of most pumps are fitted with pressure relief valves or pressure regulating valves. This valve is normally a ball and spring arrangement as illustrated in Fig. 39. When the pressure increases beyond preset limits, the ball is forced off its seat against the spring pressure and excess oil is routed back into the pump or into the crankcase, thus relieving the pressure. Excess oil pressure could otherwise force oil past seals and cause engine damage.

When the engine is stopped and oil pressure drops to zero, the spring holds the ball against its seat to prevent oil from draining back into the oil pan. This keeps the system primed for lubrication when the engine is restarted.

Normally the pump is mechanically driven by the engine from either the crankshaft or the camshaft.

Fig. 39 — Pressure Regulating Valve

Fig. 40 — Plunger Pump Installed in Engine

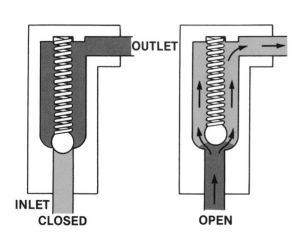

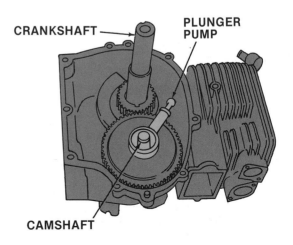

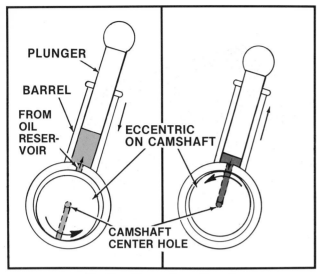

Fig. 41 — Plunger Pump Operation

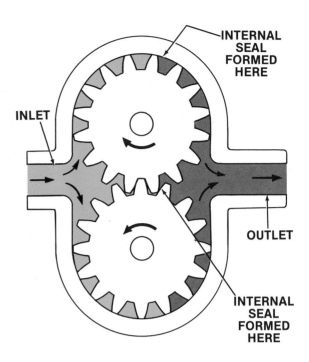

Fig. 42 — External Gear Pump

PLUNGER PUMP

The plunger pump (Fig. 40) may be mounted on the camshaft with the chamfered side of the pump body toward the camshaft gear.

The camshaft is hollow in the center with a drilled passage to the outside to align with the pump.

As the camshaft rotates, the chamfer (or eccentric) causes the barrel and plunger to stroke. As the eccentric pulls the barrel down, oil is drawn into the pump (Fig. 41, left).

As the camshaft rotates further, the inlet is blocked and the plunger moves down in the barrel, pressurizing the oil. When the drilled passage in the camshaft aligns with the pump (Fig. 41, right), pressure oil is routed through the camshaft to other engine parts for lubrication.

EXTERNAL GEAR PUMP

The **external gear pump** has two gears in mesh, closley fitted inside a housing (Fig. 42). The driveshaft drives one gear, which in turn drives the other gear. The machined surfaces of the outer housing are used to seal the gears.

As the gears rotate and come out of mesh, they trap inlet oil between the gear teeth and the housing. The trapped oil is carried around to the outlet chamber. As the gears mesh again, they form a seal which prevents oil from backing up to the inlet. The oil is forced out at the outlet and sent through the system.

ROTOR PUMP

The **rotor pump**, which is a variation on the internal gear pump, is also relatively simple in design. An inner rotor turns inside a fixed rotor ring (Fig. 43).

In operation, the inner rotor is driven inside the rotor ring. The inner rotor has one less lobe than the outer ring at any one time. This allows the other lobes to slide over the outer lobes, making a seal to prevent backup of oil.

Fig. 43 — Rotor Pump

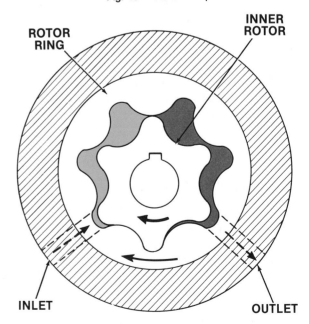

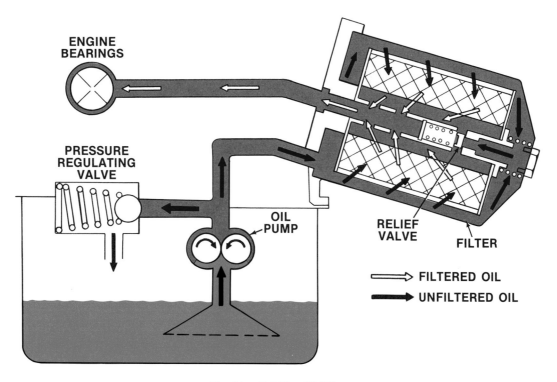

Fig. 44 — Full-Flow Oil Filter

As the lobes slide up and over the lobes on the outer ring, oil is drawn in. As the lobes fall into the ring's cavities, oil is squeezed out.

Oil Filters

Many engines are also equipped with a full-flow oil filter. This filter receives all of the oil as it leaves the oil pump and filters out contaminants before allowing the oil to reach the engine parts (Fig. 44). Oil and filters must be changed regularly at the intervals recommended by the engine manufacturer to avoid clogging the filter. Many oil filtration systems contain a filter bypass which allows oil to flow past a clogged filter and still lubricate engine parts. However, if the oil is dirty enough to clog the filter, it is too dirty to effectively lubricate engine parts.

CRANKCASE VENTILATION

During normal operation, unburned fuel vapor and water vapor are created in the engine. These vapors turn into contaminating liquids that drain into the crankcase and mix with the engine oil. To eliminate this, a crankcase breather (Fig. 45) is installed, usually over the valve lifter chamber.

Up and down movement of the piston causes air flow in and out of the crankcase. The crankcase breather allows air to flow out of the crankcase on the piston downward stroke to release the unwanted vapor. When the piston moves up, the breather blocks outside air flow into the crankcase. This creates a partial vacuum in the crankcase and prevents oil from being forced out around the piston rings, oil seals or gaskets.

ENGINE OILS

Four-cycle engine oils reduce friction and wear, cool moving parts, help seal the cylinders, and clean parts.

Fig. 45 — Crankcase Breather

SINGLE-VISCOSITY OIL   MULTI-VISCOSITY OIL

Fig. 46 — Oil Viscosity Marked on Container

To do this, they must:

- *Keep a protective film around moving parts*
- *Resist high temperatures*
- *Resist corrosion and rust*
- *Prevent ring sticking*
- *Prevent sludge formation*
- *Flow easily at low temperatures*
- *Resist foaming*
- *Resist breakdown after prolonged use*

Oil Viscosity

The oil viscosity marked on containers (Fig. 46) is the measure of fluidity of an oil at a given temperature. Lighter (thinner) oils intended for winter use carry a 0W, 5W, 10W, 15W, 20W, or 25W symbol. Heavier oils such as 30 or 40 are suggested for warm-weather use.

Some oils are compounded to exhibit qualities of light oils at cold temperatures and heavier oils at high temperatures. These oils are called multigrade or multiviscosity, such as 10W30. They are recommended for use during seasons in which both extreme cold and warm periods occur.

Most 4-cycle engine manufacturers currently recommend SAE 30 weight oil in the summer when temperatures are above 32° F (0° C) and SAE 5W-20 in the winter when temperatures are below 32° F (0° C).

Oil Performance

Oil viscosity indicates the temperature range of oil but not the performance characteristics or quality. Therefore, when selecting an oil, consider the American Petroleum Industry (API) or Military (Mil) ratings.

Most 4-cycle engine manufacturers make the following recommendations:

- *API Service Rating CD/SE, CD/SD or SD*
- *Military Service Rating Mil-L-46152 or Mil-L-2104C*

SUMMARY

Basic components common to most 4-cycle engines are:

Cylinder block — Supports the rest of the engine and holds components in proper relationship.

Cylinder — Hollow tube in which piston works and compression and power are developed.

Cylinder head — Seals one end of cylinder so power and compression can be developed.

Piston — Slides in cylinder and uses piston rings to seal compression.

Connecting rod — Links piston to crankshaft.

Crankshaft — Converts linear motion of piston, through connecting rod, into rotary motion.

Flywheel — A disk keyed to one end of the crankshaft which stores energy, stabilizes crankshaft speed, transmits power, and acts as a fan on some air-cooled engines.

Camshaft — Responsible for timing, lift, and duration of valve movement.

Valve train — Provides a means of admitting air (and fuel in gasoline engines) to the cylinder and exhausting burned gases.

Lubrication systems — Provides an oil film to engine components which rub together to reduce friction and cool the parts.

CHAPTER 2 REVEIW

1. (True or false) The cylinder block is the foundation that supports the rest of the engine.

2. What are the two basic types of cylinders?

3. (Fill in the blank) The primary purpose of the cylinder head is to seal one end of the _____ .

4. (Fill in the blanks) Gasoline and diesel cylinder heads have threaded holes from the outside of the head to the combustion chamber. The holes are for a _____ _____ or _____ _____ .

5. (True or false) Cylinder heads are designed to be air-cooled only.

6. Name the two basic types of piston rings.

7. (True or false) The lands on the piston are actually grooves which carry the piston rings.

8. (Fill in the blanks) The _____ _____ links the piston to the crankshaft.

9. Name the parts of the crankshaft.

10. Match the following terms to the correct definition below: plain bearing (insert), antifriction bearing, bushing

   1. One-piece cylindrical bearing with sliding contact between surfaces.

   2. Rollers or balls used where minimum friction is desired.

   3. Two-piece bearing with locking tabs to prevent rotation.

11. Name four possible functions of the flywheel.

12. (Select the correct term) The crankshaft operates at (half, twice) the speed of the crankshaft.

13. (Fill in the blanks) The design of the cam lobe determines the _____, _____, and _____ of valve movements.

14. Explain the function of automatic release mechanisms.

15. Identify at least five components of a valve train.

16. (Fill in the blank) The two types of lubrication systems commonly found in compact equipment engines are _____ and _____ .

17. (True or false) Pressure lubrication systems usually contain a dipper or slinger to provide oil to lubrication points.

18. (True or false) The crankcase breather allows air to flow out of the crankcase to release unwanted vapors.

19. Explain why the same engine may need to use oils of different viscosities at different times of the year.

CHAPTER 3

BASIC 2-CYCLE ENGINE

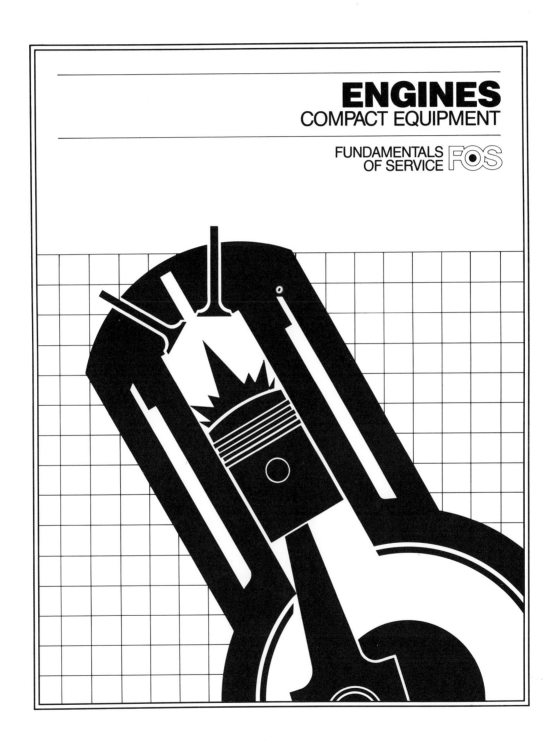

ENGINES
COMPACT EQUIPMENT

FUNDAMENTALS
OF SERVICE F⊙S

SKILLS AND KNOWLEDGE

This chapter contains basic information that will help you gain the necessary subject knowledge required of a service technician. With application of this knowledge and hands-on practice, you should learn the following:

- The components of a 2-cycle engine.

- The function of each component of a 2-cycle engine.

- How the components work.

- How lubrication is accomplished on 2-cycle engines.

- Fuel and oil requirements for 2-cycle engines.

- Why chain saws have some different operating requirements than other 2-cycle engines.

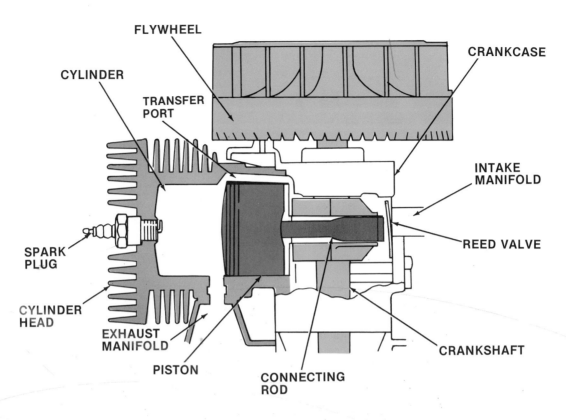

FLYWHEEL

CRANKCASE

CYLINDER

TRANSFER
PORT

INTAKE
MANIFOLD

SPARK
PLUG

REED VALVE

CYLINDER
HEAD

EXHAUST
MANIFOLD

CRANKSHAFT

PISTON

CONNECTING
ROD

Fig. 1 — Two-Cycle Engine Components

INTRODUCTION

Two-cycle engines are generally simpler in design and contain fewer components than 4-cycle engines. For example, most 2-cycle engines have no valve train.

Although 2-cycle engine operation is different than 4-cycle operation, many components are similar. However, because of the differences, similar components may have a slightly different appearance.

This chapter discusses the operation of the two basic types of 2-cycle engines found on compact equipment such as chain saws, lawn mowers, and snow throwers. They are:

- **Reed valve**

- **Piston ported**

Components common to 2-cycle engines (Fig. 1) are:

- **Cylinder head**

- **Cylinder**

- **Crankcase**

- **Piston**

- **Connecting rod**

- **Crankshaft**

- **Flywheel**

- **Reed valves (reed-valve engine)**

- **Lubrication system**

Each component will be discussed individually.

Chain saws contain these same components but also have some special operating requirements. These will be discussed at the end of the chapter.

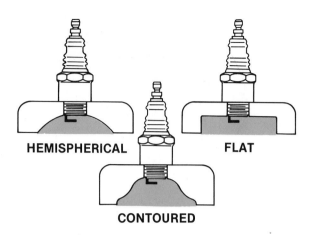

HEMISPHERICAL   FLAT

CONTOURED

Fig. 2 — Cylinder Head Designs

HEAD GASKET

Fig. 4 — Gasket "Crush" Provides Seal

CYLINDER HEAD

The primary purpose of the cylinder head is to seal one end of the cylinder so that the piston can develop compression and power. Again, because of engine requirements for spark ignition, and because of single- and multiple-cylinder designs, cylinder-head appearance varies (Fig. 2). The cylinder-head contour must be carefully matched to the requirements of the engine combustion chamber. Thus, cylinder heads of many patterns exist, but all share these common features:

- *A means of sealing tightly against the cylinder*

- *A combustion chamber*

- *Accommodation for spark ignition*

- *Provision for cooling*

CYLINDER-HEAD SEALING

The cylinder head must be tightly sealed against the cylinder to prevent leaking and a resulting loss of compression and power. This seal is usually accomplished by placing a gasket between mating flat surfaces on the cylinder and cylinder head (Fig. 3).

Fig. 5 — Combustion Chamber

Fig. 3 — Cylinder Head Gasket

HEAD GASKET

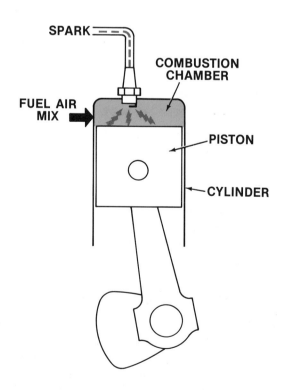

SPARK

COMBUSTION CHAMBER

FUEL AIR MIX

PISTON

CYLINDER

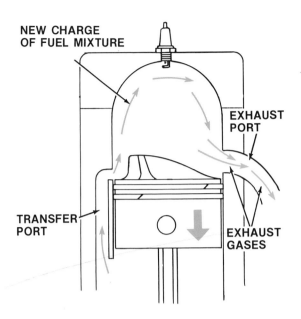

Fig. 6 — Combustion Chamber Turbulence and Exhaust

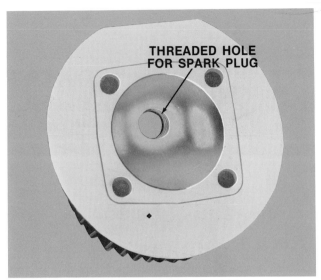

Fig. 7 — Threaded Hole for Spark Plug

When the head is bolted tightly to the cylinder, the gasket is crushed between the two surfaces and forms a tight seal (Fig. 4). This prevents the compressed gases from escaping and causes them to exert their maximum force on the piston.

COMBUSTION CHAMBER

The combustion chamber is the confined volume between the cylinder head and the top of the piston (Fig. 5). The cylinder head portion of the combustion chamber is a concave or flat area on the bottom of the head. It is shaped to:

• *Provide the controlled fuel and air flow and turbulence for proper combustion (Fig. 6)*

• *Provide a confined area in which fuel is burned and power is developed*

• *Allow the spent gases to escape through the exhaust port (Fig. 6)*

Obviously, different engine designs will affect the shape of the combustion chamber. The size (total volume) of the chamber is determined by the desired compression ratio.

COMBUSTION CHAMBER IGNITION

Cylinder heads for gasoline engines normally have a threaded hole from the outside of the head to the inside of the combustion chamber (Fig. 7). This hole is designed to accommodate a spark plug which is used to ignite the fuel-air mixture by an electrical spark.

COOLING AND LUBRICATION

Air-cooled cylinder heads contain cooling fins to transfer heat to the air (Fig. 8). If the engine uses a blower to move the air and shrouding to direct it, the cylinder-head fins are arranged to take advantage of the available air flow. Coverage in this manual will be limited to air-cooled 2-cycle engines. Liquid-cooled 2-cycle engines are generally high-performance engines found on equipment such as snowmobiles and motorcycles. Engines for these types of equipment are not covered specifically in this manual.

Fig. 8 — Cooling Fins in Cylinder Head

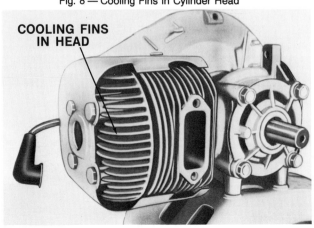

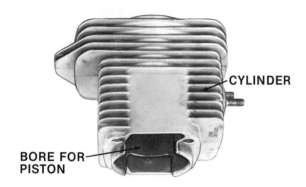

Fig. 9 — Two-Cycle Engine Cylinder

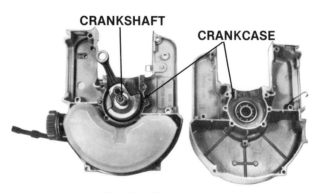

Fig. 11 — Crankcase

CYLINDER

The cylinder is a tube in which the piston works and compression and power are developed (Fig. 9). The cylinder forms part of the combustion chamber (the cylinder head also forms part of the combustion chamber as discussed earlier).

The cylinder on many 2-cycle engines is located between the cylinder head and crankcase. It may be bolted to the crankcase and removed for service or may be an integral part of the crankcase (Fig. 10).

CRANKCASE

The crankcase supports the crankshaft (Fig. 11). It also assists in transferring the fuel-air mixture from the carburetor to the combustion chamber.

As the piston moves upward on the compression stroke, a vacuum is created in the crankcase. When the piston moves far enough to uncover the intake port, the fuel-air mixture is drawn through the port into the crankcase (Fig. 12). In an engine equipped with reed

Fig. 10 — Cylinder Bolted to Crankcase (left)   Cylinder Integral with Crankcase (right)

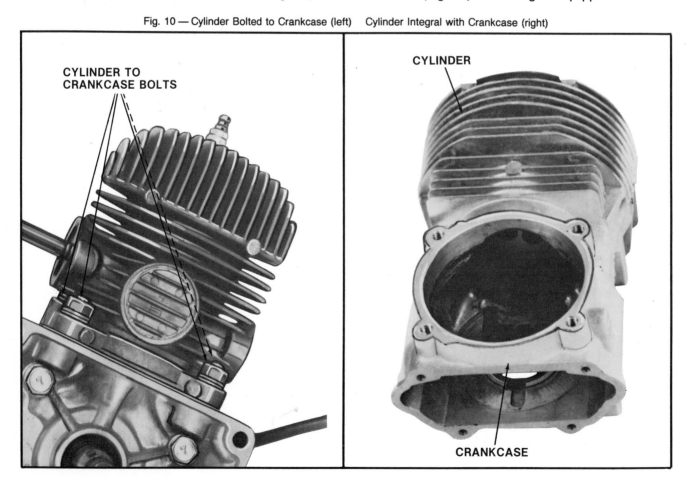

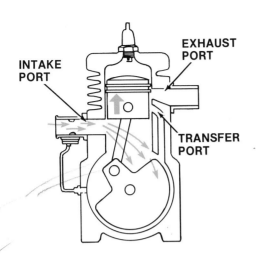

Fig. 12 — Fuel-Air Mixture Enters Crankcase

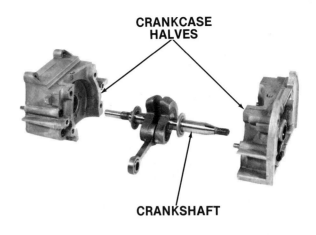

Fig. 14 — Two-Piece Crankcase

valves, the crankcase vacuum causes the reed valve to open. The fuel-air mixture passes through the reed valve and into the crankcase.

When the piston moves down on its power stroke, the intake port is again covered or the reed valve closes. The fuel-air mixture is trapped in the crankcase. Downward movement of the piston causes pressure in the crankcase to rise. When the piston clears the transfer port, the pressurized fuel-air mixture transfers to the combustion chamber (Fig. 13) and also forces out the exhaust gas.

A two-piece crankcase is common on many 2-cycle engines (Fig. 14). This makes crankshaft service and repair easier.

PISTON

The piston (Fig. 15) slides in the cylinder and is fitted closely to the cylinder to seal the compression. The piston and cylinder heat up during engine operation and may expand at different rates. Therefore, flexible members—the piston rings—are used to take up the differences.

PISTON RINGS

Piston rings are cast iron, or combinations of cast iron and steel. They fit into grooves machined into the circumference of the piston (Fig. 15). When the piston

Fig. 13 — Mixture Transferred to Combustion Chamber

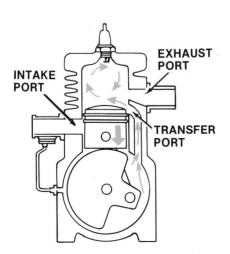

Fig. 15 — Two-Cycle Engine Piston

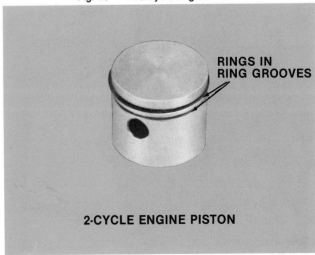

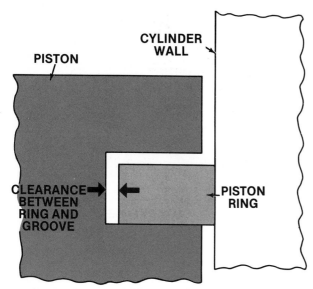

Fig. 16 — Piston Ring Clearance in Groove

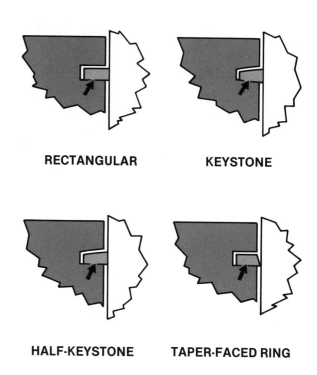

Fig. 18 — Types of Compression Rings

and rings are installed in the cylinder, the rings on a 2-cycle engine perform two functions:

• *Expand against the cylinder walls to form a gas-tight seal between piston and cylinder*

• *Help cool the piston by transferring heat*

The clearance between rings and piston grooves (Fig. 16) allows the piston to expand with little risk of scuffing the cylinder wall.

Two-cycle-engine pistons normally contain two piston rings — both compression rings. An oil control ring, found in most 4-cycle engines, is not required because 2-cycle oil is mixed with gasoline to provide lubrication.

Fig. 17 — Compression Rings Seal Combustion

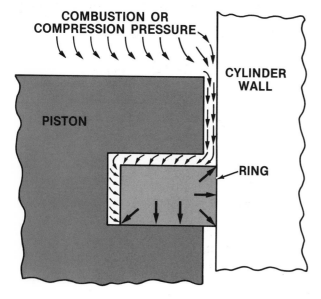

Compression rings seal the cylinder for maximum compression. They expand against the cylinder wall to provide the seal. Combustion pressure helps to assure a good seal by forcing the ring against the bottom of the ring groove and out against the cylinder wall (Fig. 17). Compression rings also remove small carbon deposits from the cylinder walls during operation.

Four types of compression rings may be used on 2-cycle engines. The type used depends on engine design.

Rectangular rings contact the cylinder wall along the entire face (Fig. 18). **Keystone and half-keystone rings** (Fig. 18) help to reduce deposits that may fill ring grooves and cause sticking. They also contact the cylinder wall along the entire face. **Taper-faced rings** (Fig. 18) contact the cylinder wall with only the lower outside edge. They seat quickly and provide good oil wiping on the downstroke.

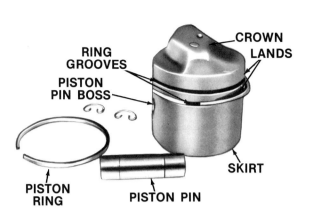

Fig. 19 — Parts of Piston Body

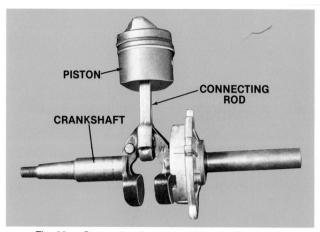

Fig. 20 — Connecting Rod Links Piston to Crankshaft

PARTS OF THE PISTON BODY

The parts of the piston body are (Fig. 19):

• **Crown** — The top surface which receives the pressure of combustion. The crown may be shaped to increase compression and swirl the charge in the manner of a turbulence chamber.

• **Ring grooves** — Carry the piston rings

• **Lands** — The raised areas between the ring grooves which aid in containing and applying pressure to the rings. They receive the force of the combustion gases as applied to the rings.

• **Piston-pin boss** — Accepts the piston pin and provides the force-bearing surfaces used to transmit power and motion to the connecting rod.

• **Piston pin** — Connects the piston to the connecting rod. The pin may rotate freely in both connecting rod ends and pin bosses (full-floating pin). Piston pins require retainers to keep them in position within the bosses.

• **Piston skirt** — The cylindrical area below the ring grooves which may be ground to a slightly elliptical and tapered shape. This shape allows the piston to expand to a cylindrical shape as it heats up. This results in a better gas seal and better fit for less contact between the skirt and the cylinder walls. Pistons used in the lighter and less expensive engine applications may be completely cylindrical instead of cam ground.

The piston must be properly weighted and balanced to provide smooth engine performance. This is especially true with multiple-cylinder configurations where the cylinders must balance each other precisely to void excessive engine vibration.

CONNECTING ROD

The connecting rod links the piston to the crankshaft (Fig. 20). The parts of the connecting rod (Fig. 21) and their function are:

• **Eye** — Accepts the piston pin and provides the direct link to the piston.

• **Head and cap** — The direct connection to the crankshaft. They are cut from a single piece head and fitted with the crankpin bearings, if bearings are used.

• **Shank** — The solid, normally tapered, link between head and eye which is cast or forged as a part of the connecting rod.

• **Cap bolts** — Used to attach the connecting rod cap to the head to secure the rod to the crankshaft.

The connecting rod eye is equipped with a bushing or bearing closely fitted to the piston pin. This bushing provides a thrust and wear surface for the pin.

The head and end cap may be fitted with a split bearing

Fig. 21 — Connecting Rod Parts

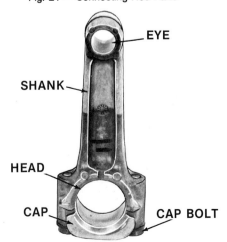

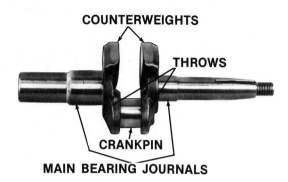

Fig. 22 — Crankshaft Parts

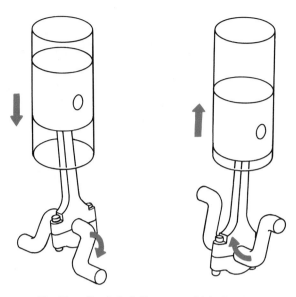

insert set which provides the thrust and wear surfaces against the crankpin. Many engines on compact equipment with lightweight metal connecting rods do not use separate bearings.

CRANKSHAFT

The crankshaft (Fig. 22) converts the linear motion of the piston, through the connecting rod, to rotary motion. Since the crankshaft is subject to severe stresses it is usually a high-strength casting or forging, precision machined, balanced, and carefully inspected. The parts of the crankshaft are:

• **Journals** — Bearing surface used to support the shaft. They are usually hardened.

• **Crankpins (connecting rod journals)** — Bearing surfaces which accept the thrust of the connecting rod. They are usually hardened.

• **Throws** — The offsets which support the crankpins and provide the leverage to rotate the crankshaft.

• **Counterweights** — Balancing weights opposite the crankpins.

Fig. 23 — Crankshaft Rotates on Main Bearings

The main-bearing journals are mounted in bearing surfaces in the front and rear walls of the crankcase. The crankshaft rotates on these fixed bearing points under the thrust of the connecting rod against the crankpin (Fig. 23).

The bearings used on the crankpins (if bearings are used) differ from the bushing used on the eye end of the connecting rod. The bushing is a one piece cylinder and the crankpin bearings are generally two piece inserts (Fig. 24). Two-cycle engines often use needle bearings around the crankpin instead of bushings or split bearings.

Antifriction bearings are often found in 2-cycle engines. Such ball, roller, or needle bearings (Fig. 25) installed on crankpins and journals, are used when minimum friction is desired. Roller bearings are polished steel cylinders. They are arranged around a crankpin or

Fig. 24 — Bushings and Plain Bearings

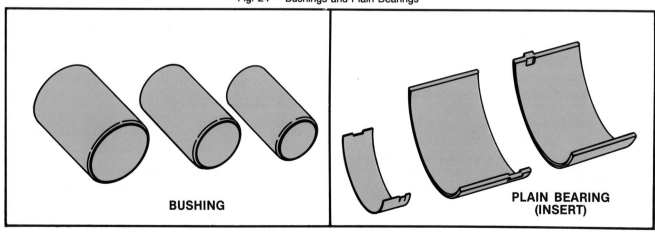

BUSHING

PLAIN BEARING (INSERT)

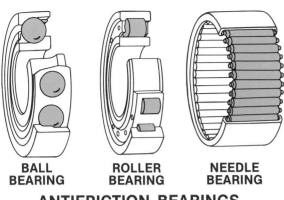

ANTIFRICTION BEARINGS

Fig. 25 — Antifriction Bearings

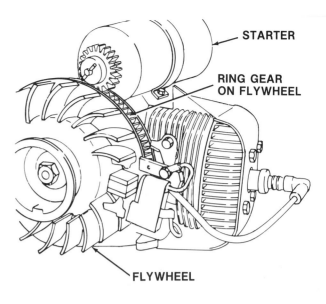

Fig. 27 — Starter Motor Turns Flywheel

bearing journal, usually in preformed races which contain the bearings. When the engine is running the motion of the bearing is rolling instead of sliding.

The crankshaft throw positions affect:

- **Engine balance**
- **Torsional vibration of the crankshaft**
- **Main-bearing loading**

Therefore, the crankshaft throws are statically and dynamically balanced to keep vibration at a minimum and main-bearing loading as constant as possible.

FLYWHEEL

The flywheel is a disk of iron, steel, or aluminum bolted and sometimes keyed to one end of the crankshaft. The design of the flywheel depends on engine design and cooling requirements (Fig. 26). It is used to:

- *Store energy between power pulses of the engine*
- *Stabilize the speed of the crankshaft*
- *Transmit power to the load*
- *Act as a fan for some air-cooled engines*

The flywheel provides the inertia to spin the engine through its compression and intake stroke without slowing down. A heavy enough flywheel will actually dampen the vibration of a single-cylinder engine so that there is little sign of the individual power pulses.

But, a heavy flywheel also has disadvantages. The amount of energy required to bring it to speed increases with the weight of the flywheel. Also, a heavy flywheel hampers the ability of the engine to change speed quickly.

Other common functions of the flywheel are:

- **Starting device (electric-start engines)** — A gear on the starter motor meshes with a ring gear on the flywheel. The starter motor spins the flywheel (Fig. 27) and crankshaft until the engine starts.

- **Power line** — When fitted with a clutch and pressure plate, the flywheel is used as a direct link in the driveline.

Fig. 26 — Flywheel with Cooling Fins

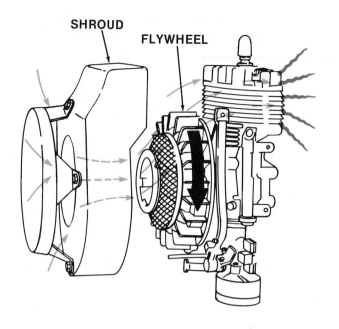

SHROUD

FLYWHEEL

Fig. 28 — Air Flow for Engine Cooling

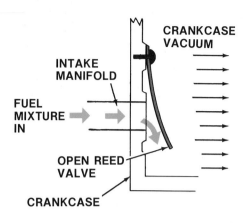

CRANKCASE VACUUM

INTAKE MANIFOLD

FUEL MIXTURE IN

OPEN REED VALVE

CRANKCASE

Fig. 29 — Reed Valve Operation

• **Engine cooling** — Most air-cooled engines of the type used on compact power equipment use fins on the flywheel to act as a blower for cooling air when the engine is running (Fig. 28).

• **Electrical** — The magneto ignitions on many engines use one or more magnets cast into the flywheel as a part of the ignition current and battery charging generating system.

The size and weight of the flywheel used depends on the amount of vibration and pulse dampening needed. Other considerations are the number of cylinders and compression ratio of the engines, the number and types of extra systems for which the flywheel drives or supplies power, the speed response required, and the size and type of load the engine is expected to drive.

REED VALVES

Reed valves provide a means for allowing a fuel-air mixture to enter the crankcase on the compression stroke (Fig. 29).

One end of the flexible, spring steel reed valve is fastened while the other end is free to flex. A reed stop keeps the reed from flexing too far and becoming permanently bent.

Not all 2-cycle engines contain reed valves. Those which do not are called piston-ported engines. Different engine designs determine the need for reed valves.

LUBRICATION SYSTEMS

Two-cycle engines normally have one of two types of lubrication systems:

• **Fuel-oil premix**

• **Oil injection**

Both systems accomplish four basic functions:

• *Reduce friction caused by metal-to-metal contact between moving parts*

• *Seal the cylinder to obtain optimum combustion pressure*

• *Keep the engine clean by cleaning carbon from between moving parts and by preventing the formation of deposits*

• *Cushion shock loads to reduce noise and wear*

Coverage in this manual will be limited to fuel-oil premix lubrication. Oil injection is generally found on high-performance engines on snowmobiles and motorcycles. Engines for those types of equipment are not covered specifically in this manual.

Fig. 30 — Fuel and Oil Mixed in Container

Fuel-Oil Premix

In these types of engines, gasoline and 2-cycle engine oil are mixed together in a container and then transferred to the fuel tank (Fig. 30). Follow the manufacturer's recommendation for the correct fuel-oil mixture and mix as follows:

1. Pour half the gasoline into a clean container

2. Pour in all the required oil

3. Shake mixture vigorously

4. Pour in remaining gasoline

5. Shake mixture vigorously

6. Mark the container to prevent it from being confused with a container of some other liquid or mixture.

Check manufacturer recommendations for type of gasoline specified for the engine (regular, unleaded, or premium) as well as type of 2-cycle engine oil.

Use only oil refined especially for 2-cycle engines. These oils are identified by the Boating Industry Association (BIA) with the 2-cycle, water-cooled (TCW) rating marked on the can.

Two-cycle oil with this rating contains:

• *Coupling agents to improve mixability with gasoline even at low temperatures.*

• *Detergents to prevent varnish buildup and ring sticking.*

• *Antioxidants to raise the temperature at which the oil will burn.*

• *Rust inhibitors to protect the cylinder, piston, rings, and bearing from rust.*

• *Antiwear agents to prevent scuffing and piston seizure.*

Also, check the operator's manual for the correct gasoline-to-oil mixture ratio (usually about 50:1). Use a clean container to mix the fuel.

The fuel-oil mixture in these engines may be gravity fed or pumped by a fuel pump from the fuel tank to the carburetor. After being metered through the carburetor, the fuel-oil mixture is drawn into the engine for lubrication and combustion.

The amount of oil in the fuel-oil mixture is adequate for lubricating the crankshaft bearings, piston pin bushing, piston rings, and the cylinder walls. The flow of the fuel-oil mixture through the crankcase and combustion chamber routes the oil to these lubrication points.

Fig. 31 — Chain Saws Operate in Many Positions

CHAIN SAW ENGINES

Chain saw engines have different operating requirements than most other 2-cycle engines. Chain saw engines must:

• *Operate well in all attitudes and positions (Fig. 31)*

• *Be light and compact while maintaining a high level of performance*

• *Be extremely responsive to throttle variations*

• *Be able to operate under adverse conditions*

Fig. 32 — Crankcase is Part of Saw Main Frame

**FRAME OF
CHAIN SAW**

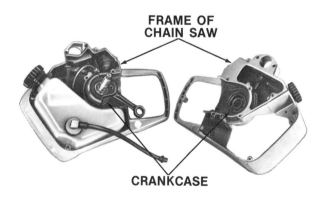

CRANKCASE

FUEL DELIVERY

For a chain saw to operate well in any position, a special diaphragm type carburetor is used. (Notable exceptions are the huge timber-cutting saws which may be equipped with float type carburetors. They are generally operated within a few degrees of upright at all times.) This type of carburetor will be discussed in Chapter 4 along with other fuel system requirements for chain saws.

CRANKCASE

The crankcase of a chain saw engine is compact and designed as the main frame of the saw as well (Fig. 32). The crankshaft is well sealed and machined to be fitted with the clutch. The clutch controls the movement of the chain. Because of this fitted design, all of the components of the chain saw fit into a tight, compact package. Components are well shielded from dirt and other outside contaminants by a close-fitting shroud. The shroud also directs cooling air blown in by the fins on the flywheel.

SUMMARY

Some basic components common to most 2-cycle engines are:

Cylinder head — Seals one end of cylinder so power and compression can be developed.

Cylinder — Hollow tube in which piston works and compressions and power are developed.

Crankcase — Supports the crankshaft and assists in transferring fuel-aix mixture to combustion chamber.

Piston — Slides in cylinder and uses piston rings to seal compression.

Connecting rod — Links piston to crankshaft.

Crankshaft — Converts linear motion of piston, through connecting rod, into rotary motion.

Flywheel — A disk keyed to one end of the crankshaft which stores energy, stabilizes crankshaft speed, transmits power, and acts as a fan on some air-cooled engines.

Reed valve — A flexible, spring steel reed acting as a means to allow the fuel-air mixture to enter the crankcase.

Lubrication system — Provides an oil film to engine components which rub together to reduce friction and cool the parts.

Two-cycle engine oil is usually mixed with gasoline before placing in fuel tank (fuel-oil premix lubrication).

CHAPTER 3 REVIEW

1. What are the two basic types of 2-cycle engines?

2. (Multiple choice) The primary purpose of the cylinder head on a 2-cycle engine is to:

    A. seal one end of the cylinder for compression and power

    B. provide a location for the valve train

    C. give the engine the proper balance

    D. act as a foundation to support the rest of the engine

3. (Fill in the blank) Air-cooled cylinder heads contain cooling _____ to transfer heat to the air.

4. (True or false) Cylinders on all 2-cycle engines are bolted to the crankcase.

5. (True or false) Two-cycle engines contain two types of piston rings — compression and oil control.

6. (Fill in the blanks) The connecting rod _____ accepts the piston pin while the _____ and_____ connect the rod to the crankshaft.

7. What is one disadvantage of a heavy flywheel?

8. (True or false) Two-cycle engines which do not contain reed valves are known as piston-ported engines.

9. Name the two types of lubrication systems found on 2-cycle engines.

10. Chain saw engines have different operating requirements than most other 2-cycle engines. What are some of these special requirements?

CHAPTER 4

GASOLINE AND DIESEL FUEL SYSTEMS

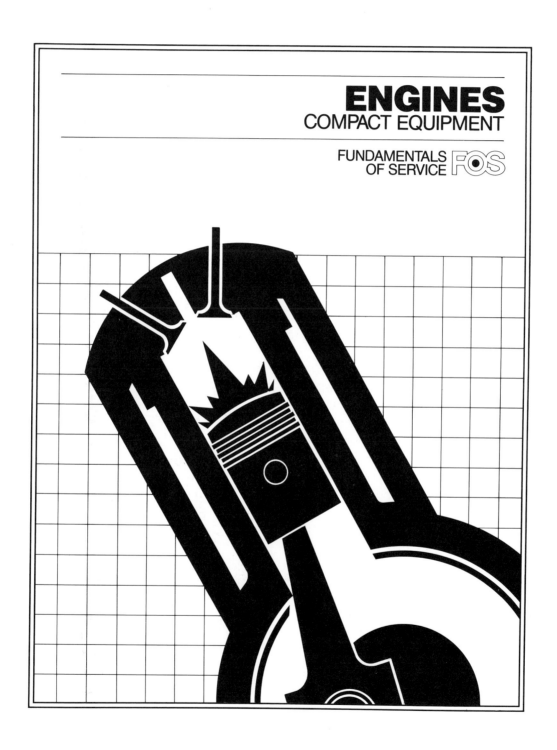

ENGINES
COMPACT EQUIPMENT

FUNDAMENTALS
OF SERVICE FOS

SKILLS AND KNOWLEDGE

This chapter contains basic information that will help you gain the necessary subject knowledge required of a service technician. With application of this knowledge and hands-on practice, you should learn the following:

Gasoline Fuel System

• Components of the system

• Two types of systems

• How the fuel system works

• Operation of each component

• Types of carburetors and how each works

• Types of governors and operation

Diesel Fuel System

• Components of the system

• How the fuel system works

• Operation of each component

• Basic requirements of fuel injection

• Types of injection pumps and operation

• How the turbocharger works

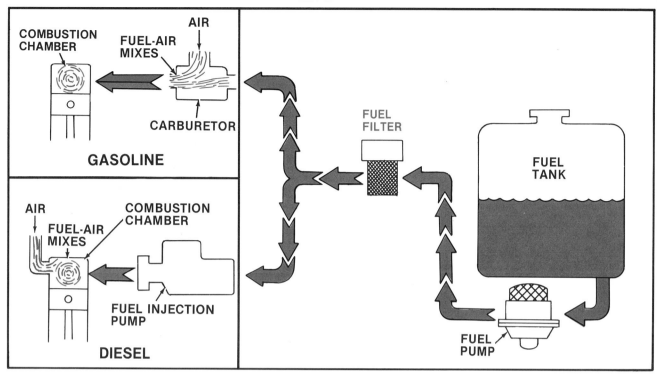

Fig. 1 — Basic Fuel System

The fuel system (Fig. 1) is responsible for:

- *Storing fuel to be used by the engine*

- *Transferring fuel from the tank to the engine*

- *Filtering fuel and air*

- *Metering fuel and air*

- *Exhausting burned fuel from the engine*

This chapter will cover the components involved in each of these functions and the operation of each component. System maintenance and repair will be covered in Chapter 7.

TYPES OF FUEL SYSTEMS

Two basic types of fuel systems are found on compact equipment such as lawn mowers, lawn and garden

Fig. 2 — Equipment Having Different Types of Fuel Systems

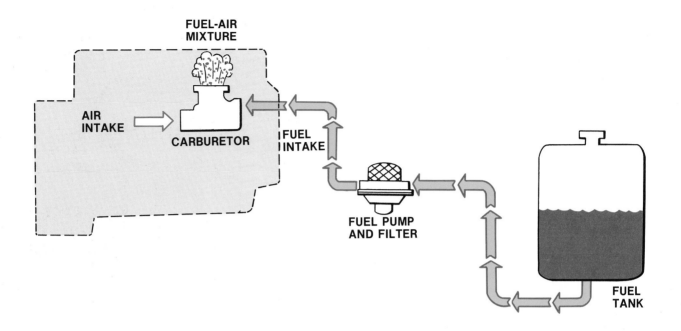

Fig. 3 — Gasoline Fuel System

tractors, compact utility tractors, and skid-steer loaders (Fig. 2). They are:

- **Gasoline**

- **Diesel**

The system used depends on the engine design. This chapter will discuss the components and operation of both systems. The discussion will consider the air intake and exhaust systems as part of the fuel system.

Fig. 4 — Gravity-Feed and Force-Feed Systems

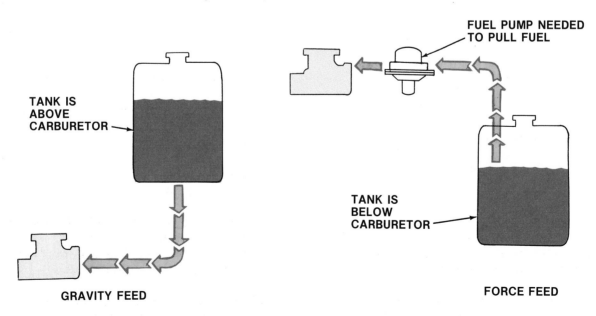

The first part of the chapter will cover gasoline systems and the last part will cover diesel systems.

GASOLINE FUEL SYSTEMS

Gasoline fuel systems generally consist of the following components (Fig. 3).

- **Fuel tank**
- **Fuel lines**
- **Fuel pump**
- **Fuel filter**
- **Air cleaner (not shown)**
- **Carburetor**
- **Governor (not shown)**

The operation of each component will be discussed later. First, let's examine two methods of transferring fuel from the fuel tank to the engine. They are:

- **Gravity feed**
- **Force Feed**

GRAVITY-FEED SYSTEM

The gravity-feed system (Fig. 4, left) has a fuel tank placed above the carburetor; fuel lines; fuel filter; and the gravity-feed carburetor.

A float may be attached to a valve which allows fuel to enter the carburetor at the same rate at which the engine is consuming it. This system maintains a uniform level in the carburetor regardless of the amount of fuel in the tank.

FORCE-FEED SYSTEM

The force-feed system allows the fuel tank to be located at a level below the carburetor (Fig. 4, right).

A fuel pump is required for this system to raise the fuel from the tank to the carburetor. Fuel pump operation is described later in this chapter.

SYSTEM OPERATION

The basic force-feed system operation is shown in Fig. 5. The gravity-feed system will operate in much the same manner, except no fuel pump is used.

The *fuel pump* draws the gasoline through a fuel line from the tank and forces it to the *float chamber or bowl* of the carburetor, where it is stopped.

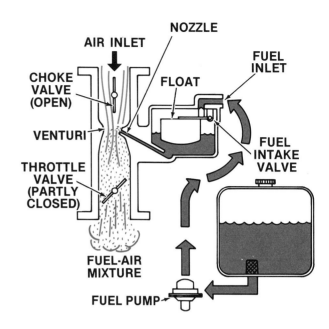

Fig. 5 — Force-Feed System Operation

The *carburetor* is basically an air tube which operates by a differential in air pressure. The carburetor will be covered in detail later in this chapter.

Filtered air is drawn into one end of the carburetor, mixes with the gasoline, and fuel-air vapor is released at the other end.

The pressure differential is created when air flows through the narrow neck, called the *venturi*. Air flow moves faster through a restriction and this lowers the air pressure.

At the same time, the engine creates a partial vacuum on intake stroke and this causes vapor to be drawn into the combustion chamber of the cylinder.

The fuel is forced into the airstream from the *nozzle* which projects into the tube at the venturi. As low-pressure air rushes by, small drops of fuel are forced out and mixed with the air.

The fuel-air mixture must pass the *throttle valve* which opens or closes to let the correct volume of mixture into the engine. This is controlled by the operator by a lever or pedal and determines engine speed.

The *choke* is a valve which controls the supply of fuel-air mixture to the engine. When starting the engine in cold weather, for example, it can be partly closed.

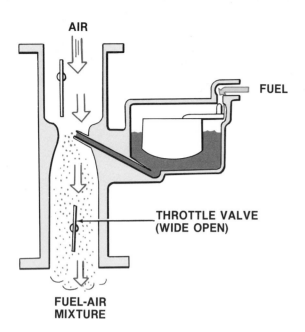

Fig. 6 — Fuel Flow at Full Throttle

forming a restriction. This restriction causes more fuel and less air to be drawn into the combustion chambers. This results in a richer mixture in the cylinders and promotes faster starting.

Fig. 5, shows the operation of the carburetor at part throttle or low power, while Fig. 6 shows operation at full throttle or full power.

Actual carburetors are more complex and will be covered later in this chapter.

GASOLINE FUEL SYSTEM COMPONENTS

Each component of the gasoline fuel system (listed earlier) has a specific function.

FUEL TANK

The fuel tank may be made of sheet metal or of a plastic or synthetic material (Fig. 7). Fuel tank capacity will vary depending on the size of the engine and the machine.

Most systems have the fuel line attached at or near the bottom of the tank and usually have a filter screen at the fuel line connection.

A drain cock may be present at the bottom of the tank to allow water and sediment to be drained periodically. A shut-off valve may be used to close the fuel outlet before removing the tank.

To ventilate the fuel tank, a vent mechanism may be built into the filler cap or as a separate opening near the top. The vent allows air to replace the fuel as it is drawn out and prevents restriction of the fuel flow or a vacuum in the tank.

Fig. 7 — Fuel Tanks — Attached to Engine (left), Remote Location (right)

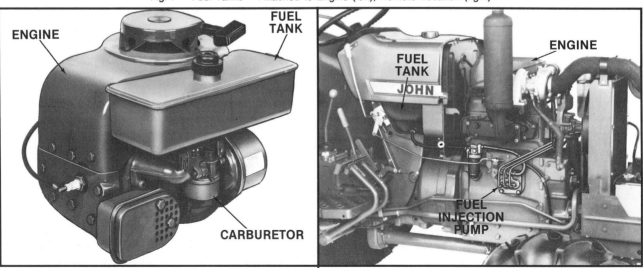

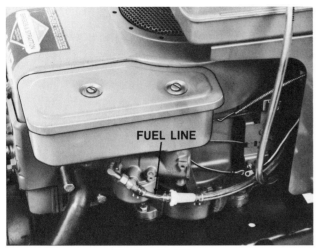

Fig. 8 — Fuel Lines

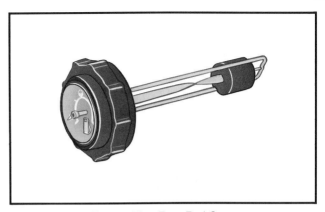

Fig. 9 — Float-Type Fuel Gauge

FUEL LINES

Fuel lines are generally made of steel tubing. However, polyethylene lines have become very popular where temperatures permit their safe use (Fig. 8).

The fuel lines simply transfer the fuel from one location to another.

FUEL GAUGES

Compact equipment may use one of two types of fuel gauges:

- **Float type**
- **Electric**

The **float-type gauge** (Fig. 9) is usually located on the fuel tank filler cap. As the level of fuel in the tank lowers, the float also lowers (Fig. 10). The relative level of the float is indicated by a pointer on the filler cap.

Fig. 10 — Float-Type Gauge Operation

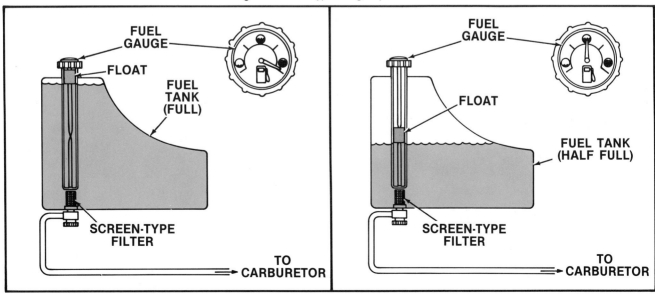

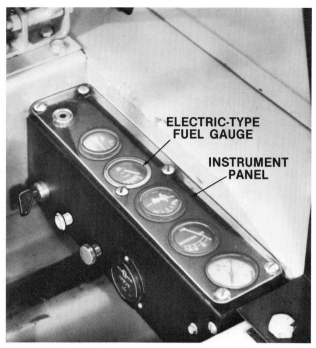

Fig. 11 — Electric-Type Fuel Gauge — Part of Instrument Panel

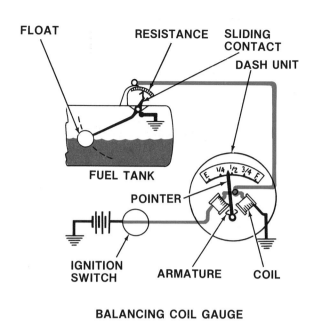

BALANCING COIL GAUGE

Fig. 12 — Electric-Type Gauge Operation

The **electric-type fuel gauge** (Fig. 11) also uses a float assembly inside the fuel tank. This float is attached to an electric unit on the tank which is wired to the fuel gauge, usually located at the operator's station. As the float follows the level of fuel in the tank, a signal is sent to the gauge to indicate the amount of fuel in the tank (Fig. 12).

FUEL PUMPS

The fuel pump transfers the required amount of fuel from the fuel tank to the carburetor. Fuel pumps are not required for gravity-feed fuel systems.

An **electric-type** fuel pump is operated by a coil that is actuated by the alternator (Fig. 13). When the coil is activated, the plunger valve is drawn toward the plunger spring (Fig. 13, left) and closes the foot valve. Fuel flows around the plunger valve.

When the transistor cuts power to the coil, the plunger spring forces the plunger valve back in its bore (Fig. 13, right). This forces the fuel around the plunger valve out of the pump to the carburetor.

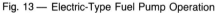

Fig. 13 — Electric-Type Fuel Pump Operation

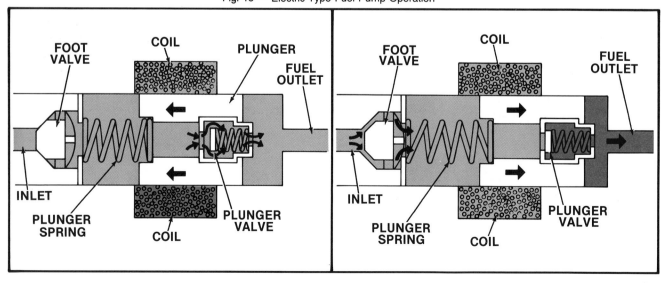

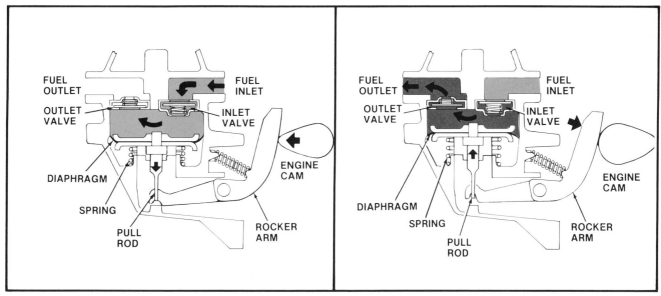

Fig. 14 — Diaphragm — Type Fuel Pump Operation

Most **diaphragm-type** fuel pumps found on compact equipment 4-cycle engines, are operated by a rocker arm which rides on a cam lobe on the engine camshaft (Fig. 14, left). As the camshaft turns, the cam lobe causes the rocker arm to rock back and forth. The inner end of the rocker arm is linked to a diaphragm located in the pump housing. As the rocker arm rocks, it pulls the diaphragm down, then releases it. A spring, under the diaphragm, forces it back up. Thus the diaphragm moves up and down as the rocker arm rocks.

When the diaphragm is released, the spring forces it back up, causing pressure to build up in the area above the diaphragm. This pressure closes the inlet valve and

opens the outlet valve (Fig. 14, right), forcing fuel from the pump through the outlet to the carburetor.

If the needle valve in the float bowl of the carburetor closes the inlet to the carburetor, no fuel can enter the carburetor. The fuel pump, therefore, can no longer deliver fuel.

In this case, the rocker arm continues to rock because it remains in contact with the rotating cam lobe. Back pressure from the carburetor causes the diaphragm to remain at its lower limit of travel (Fig. 15). The pressure is sufficient to overcome the diaphragm spring tension so the spring cannot force the diaphragm up. Normal operation of the pump resumes as soon as the needle valve in the float bowl opens the inlet valve, allowing the spring to force the diaphragm up.

Many 2-cycle engines have gravity-feed fuel systems and do not require a fuel pump. Those that require a pump, such as chain saw engines, usually incorporate a diaphragm-type pump as part of the carburetor. This will be covered later in the discussion on carburetors.

Fig. 15 — Fuel Pump when Carburetor Inlet is Closed

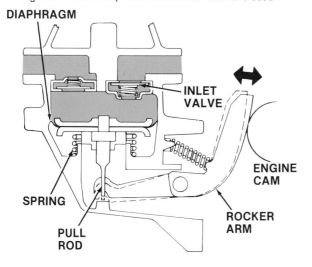

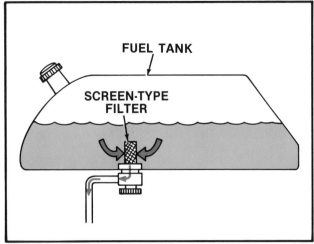

Fig. 16 — Screen-Type Fuel Screen

FUEL FILTER OR SCREEN

The fuel filter is used to remove contaminants from the fuel before the fuel reaches the carburetor. Water and dirt are contaminants which could damage the engine and affect carburetor operation. A screen-type filter (Fig. 16) will trap much of the dirt that would enter the engine. This filter could be located at any point in the fuel system between the fuel tank and the carburetor.

AIR CLEANER

Air mixes with gasoline at the carburetor to provide the engine a proper mixture for combustion. Like the fuel supply, the air must be free of contamination or the

engine could be damaged. To remove this contamination, engines are equipped with an air cleaner.

Most engines on compact equipment such as lawn mowers, lawn and garden tractors, and chain saws use one of two types of air cleaners:

- **Dry-element type**
- **Oiled-element type**

Dry-Element Type Air Cleaner

Two types of dry-element type air cleaners are usually found on compact equipment. They are:

- *Pleated-paper type*
- *Flat-element type*

In the **pleated-paper element** (Fig. 17), the first stage (precleaning) directs the air into the cleaner. The air enters at high speed so that it sets up centrifugal rotation (cyclone action) around the filter element.

The cleaner shown in Fig. 17A directs the air into the *precleaner* so it strikes one side of the metal shield. This starts the centrifugal action which continues until it reaches the far end of the cleaner housing. At this point, the dirt is collected into a dust cap, or dust unloader, at the bottom of the housing.

The cleaner shown in Fig. 17B conducts the air past *tilted fins* which start the centrifugal (cyclone) action. When the air reaches the end of the cleaner housing,

Fig. 17 — Operation — Two Types of Pleated-Paper Type Air Cleaners

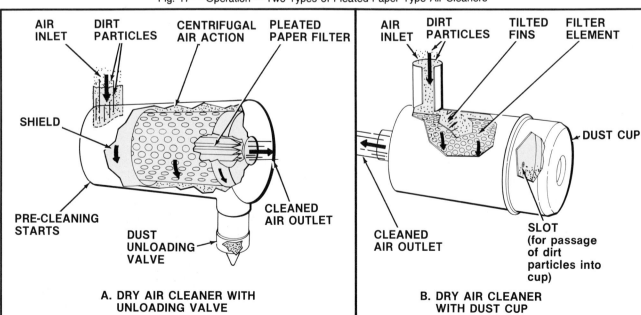

A. DRY AIR CLEANER WITH UNLOADING VALVE

B. DRY AIR CLEANER WITH DUST CUP

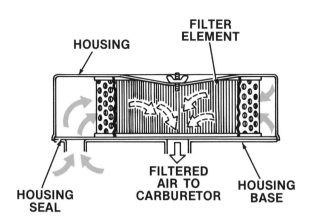

Fig. 18 — Pleated-Paper Type Air Cleaner

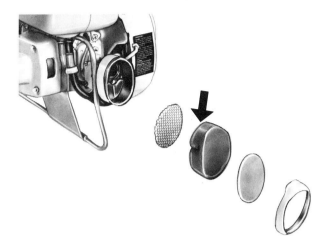

Fig. 20 — Oiled-Element Type Air Cleaner

the dirt passes through a slot in the top of the cleaner and enters the dust cup.

In both types, this precleaning action removes from 80 to 90 percent of the dirt particles and greatly reduces the load on the filter.

The partially cleaned air then passes through the holes in the metal jacket surrounding the pleated-paper filter. Filtering is done as the air passes through the paper filter. It filters out almost all of the remaining small particles. This is the second stage of cleaning.

Another type of pleated-paper type air cleaner (Fig. 18) allows air to pass directly through the paper filter. Dirt is trapped on the outside of the paper element and clean air passes through.

The **flat-element type** air cleaner (Fig. 19) is generally used on smaller compact equipment such as chain saws, weed trimmers, and some lawn mowers. The single flat element allows air to pass through while filtering out much of the airborne contamination.

Oiled-Element Air Cleaner

The oiled-element air cleaner consists of a single element (usually polyurethane foam) which fits inside of a housing (Fig. 20). The element is soaked with about five teaspoons of oil and squeezed to distribute the oil evenly and remove any excess oil.

Air passes through the element and most of the airborne dust and contaminants are trapped in the element. Most contaminants adhere more readily to the oil in the element than if the element were dry.

Fig. 21 — Carburetor Mounted on Engine

Fig. 19 — Flat-Element Type Air Cleaner

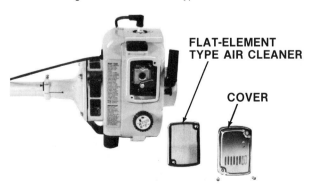

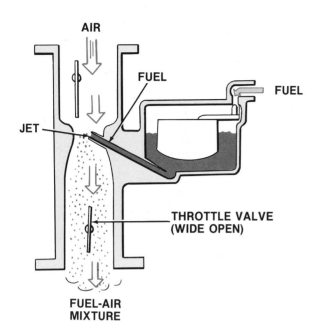

Fig. 22 — Movement of Fuel-Air Mixture

CARBURETORS

Engines will not run on liquid gasoline. The gasoline must be vaporized and mixed with air for all types of conditions. For example:

- **Cold or Hot Starting**
- **Idling**
- **Part Throttle**
- **Acceleration**
- **High Speed Operation**

By mixing fuel with air for each of these conditions, the carburetor (Fig. 21) regulates the combustion and, the power of the engine.

To get the correct fuel-air mix, the carburetor must atomize the fuel and mix fine particles of fuel with air.

Atomizing is done by adding air to the liquid fuel as it moves through the carburetor passages. This fuel-air mixture is sprayed through nozzles or jets into a stream of moving air flowing into the engine's intake manifold (Fig. 22).

The fuel in the fuel-air mixture is then vaporized before it enters the combustion chamber of the engine.

The various speed and load conditions demand a different volume of air for the mixture. The fuel-air ratio must be kept within flammable limits to permit combustion.

The primary job of the carburetor is to produce this ratio or fuel-air mix for any operating condition.

BASIC CARBURETOR PARTS

Carburetor designs vary but most share many common parts (Fig. 23):

Throat — The tube through which air is drawn into the engine and where fuel-air mixing occurs.

Venturi — A constriction in the throat which causes an increase in the speed of the air flow and, consequently, a pressure differential with the point of lowest pressure being within the venturi.

Throttle valve — A means of controlling the amount of air flow through the throat and, in some cases, the amount of fuel flow as well.

Fig. 23 — Parts Common to Most Carburetors

Fig. 24 — Float Controls Fuel Level

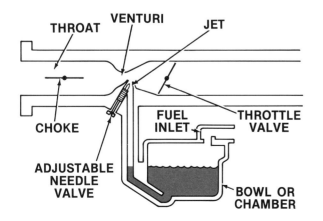

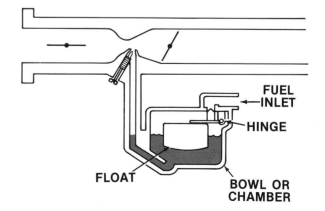

Bowl or chamber — A cavity within the carburetor where the fuel about to be used is stored. As fuel is drawn from the bowl, it is replaced so there is always a supply to respond to throttle variations. This allows the engine to run smoothly without the pulses in power output which would be caused by operating directly from a fuel pump.

Jet — A small orifice, sometimes adjustable by a needle valve, through which fuel is forced by atmospheric pressure. The jet delivers fuel in a fine spray into the carburetor throat where it mixes with air.

Choke — A device for closing off the carburetor throat at the outside of the venturi. This restricts the air flow and causes a very rich fuel-air mixture which is useful for cold starting.

TYPES OF CARBURETORS

The types of carburetors common to small gasoline engines are:

• **Float type — Adjustable and nonadjustable**

• **Suction-lift type — vacuum- and pulsating-lift types**

• **Diaphragm type**

Many variations exist among these types. The following discussion will give the general features and operation of each type.

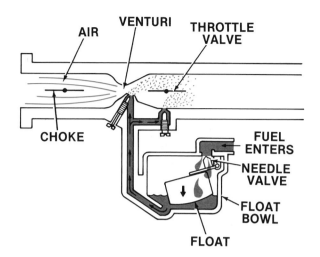

Fig. 25 — Float Valve Opens Fuel Inlet

Float-Type Carburetor (Adjustable)

Float-type carburetors get their name from the float which controls the level of the fuel in the carburetor bowl (Fig. 24). The float is constructed of lightweight material and floats on the surface of the gasoline in the float bowl. A needle valve is attached to the float. The tip of the needle valve rests in a valve seat which doubles as the orifice through which fuel enters the float bowl.

When the bowl is full the float is at its maximum height and the needle valve is forced into its seat, blocking the flow of fuel. As the fuel in the float bowl is consumed the level of the float drops with the level of gasoline and the needle valve is pulled from its seat. More fuel then is admitted to the float bowl (Fig. 25). During engine operation the needle valve may open and close as often as two hundred times per minute.

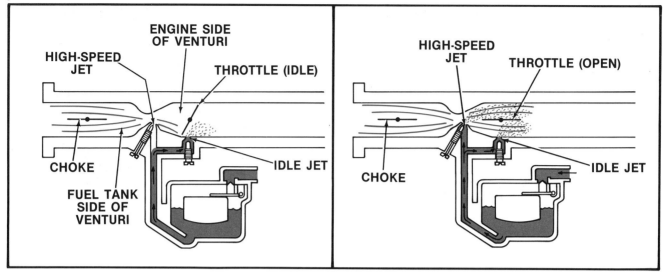

Fig. 26 — Throttle Valve and Choke Positions

The **throttle valve** in this type of carburetor is a disk, called a butterfly, mounted to pivot on a shaft inside the throat of the carburetor (engine side of the venturi)(Fig. 26). When the disk is pivoted to nearly block the carburetor throat (Fig. 26, left), a minimum of air flow is available. The engine operates at low speed and power and the throttle is said to be closed. As the disk pivots to open the carburetor throat (Fig. 26, right), more air and fuel flow and engine speed and power increase.

The **choke** is another pivoting disk (or sometimes a sliding member) mounted in the throat of the carburetor (fuel tank side of the venturi)(Fig. 26). It is used during cold starting the engine. Its operation will be covered in the next section.

OPERATION

Cold start — During the intake stroke a partial vacuum is created in the carburetor throat. When the choke is closed (Fig. 27), fuel is drawn from the high-speed and idle jets and mixes with the small amount of air which flows past the choke plate. This rich mixture is drawn into the engine and provides easier cold starting.

Idle — At idle the throttle valve is barely open. Air flow through the venturi is not fast enough to create a sufficient vacuum for the main jet to operate. Fuel then flows through the idle jet (Fig. 28). If the engine is equipped with multiple idle ports, fuel flows through the

Fig. 28 — Idle Operation

Fig. 27 — Cold Start Operation

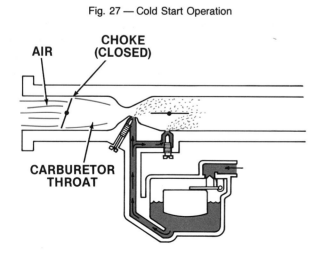

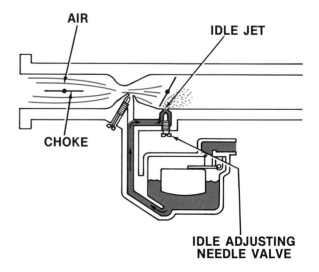

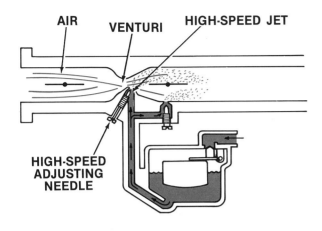

Fig. 29 — High Speed Operation

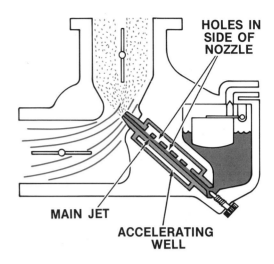

Fig. 31 — Operation at Engine Operating Speed

primary idle port when the throttle valve is in the idle position. The amount of fuel flow is controlled by the idle adjusting needle valve.

Intermediate speed — The throttle valve opens further to expose the secondary idle ports (if so equipped). Air flow through the venturi is increased and more fuel is moved. The mixture is still determined by the idle adjust valve. If the carburetor contains no secondary idle port, the main jet design will allow partial flow through the main jet at this venturi air speed.

High speed — At high speed the throttle valve is fully open (Fig. 29) and the air flow is restricted only by the

venturi. At this air flow, fuel is drawn from the high-speed jet into the venturi, regulated by the setting of the high speed mixture adjustment needle valve. Because the secondary venturi effect of the throttle is minimal at this setting, little fuel flows from the idle jet.

Float-Type Carburetor (Nonadjustable)

The nonadjustable float-type carburetor operates in the same manner as the adjustable. The difference is that all fuel metering is done internally and is designed into the carburetor. There are no adjustments available and, in some cases, no choke.

OPERATION

The rich charge needed for **cold start** is provided by a fuel well surrounding the main jet (Fig. 30). The air bleed cannot act on the float chamber until this well is cleared and, consequently, the fuel from the well enters the engine first. The rich mixture makes starting easier.

With the increased air flow resulting from the engine running at operating speed, fuel is drawn from the main jet (Fig. 31) before the well can fill. The well eventually fills so that extra fuel is available if there is a sudden demand for more power.

Fig. 30 — Operation of Nonadjustable Carburetor

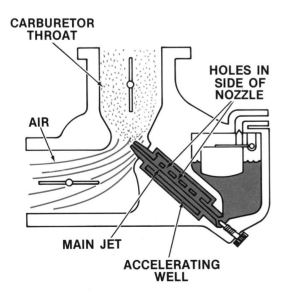

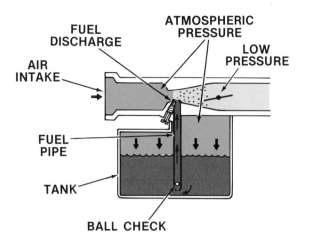

Fig. 32 — Vacuum Suction-Lift Carburetor

Suction-Lift Carburetor (Vacuum)

The vacuum suction-lift carburetor usually mounts directly on top of the fuel tank (Fig. 32). A fuel pipe, equipped with a filter screen at the pick-up end, extends down into the fuel tank. The other end of the fuel pipe is pressed or threaded into a passage in the carburetor. The high speed mixture needle valve is also a part of this pipe. The outlet of this passage consists of two discharge holes for high speed and idle operation.

The throttle valve, if one is used, may be the butterfly type as described earlier under float-type carburetors. The discharge holes open into the carburetor throat adjacent to the throttle as with the float type.

OPERATION

The intake stroke of the engine causes a pressure differential between the fuel in the tank and the fuel discharge holes in the venturi. The atmospheric pressure, acting on the fuel in the tank, forces it up the fuel pipe, past a ball check, and through the needle valve and fuel discharge holes (jets) (Fig. 33, left). The fuel is atomized, mixed with air in the venturi, and forced into the combustion chamber.

At the end of the intake stroke the pressure differential is no longer present and the ball check falls against its seat blocking the return of fuel to the fuel tank (Fig. 33, right). This causes fuel to remain in the fuel pipe at all times and increases the operating efficiency of the carburetor.

Cold start is accomplished by again using a disk-type choke. When the choke valve blocks the passage in the carburetor throat (Fig. 34), the pressure differential between carburetor throat and fuel tank causes a rich fuel-air mix to enter the combustion chamber.

At **idle**, the air flow through the carburetor is too slow to draw fuel through the high speed discharge hole. The idle discharge hole, which has no needle valve adjustment, supplies fuel during idle operation (Fig. 35). At **high speed** air flow through the carburetor throat is at maximum. The speed of the air flow is sufficient to draw fuel through the high speed discharge hole (Fig. 36).

Fig. 33 — Vacuum Suction-Lift Carburetor Operation

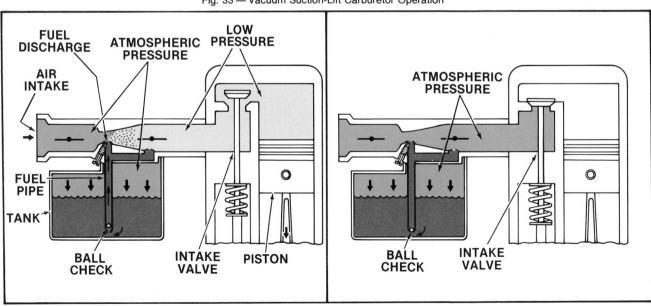

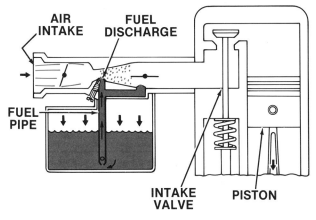

Fig. 34 — Choke Operation

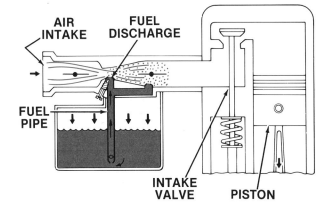

Fig. 36 — High Speed Operation

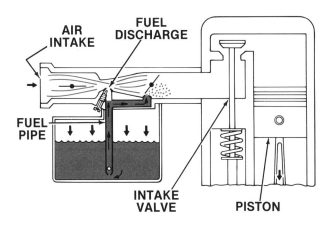

Fig. 35 — Idle Operation

Suction-Lift Carburetor (Pulsating)

The pulsating suction-lift carburetor contains a diaphragm-type fuel pump that is not found on the vacuum suction-lift type. The pump operates by reacting to pressure changes in the engine intake (Fig. 37). The pump is connected to a fuel tube from the fuel tank.

OPERATION

When the piston is on its downward stroke, the vacuum created causes fuel to be drawn from the tank through the inlet valve and into the pump chamber (Fig. 37, left). When the piston moves upward, intake manifold pressure forces the diaphragm down (Fig. 37, right). This causes valves within the pump to open and transfer fuel to the fuel reservoir.

Fig. 37 — Pulsating Suction-Lift Carburetor Operation

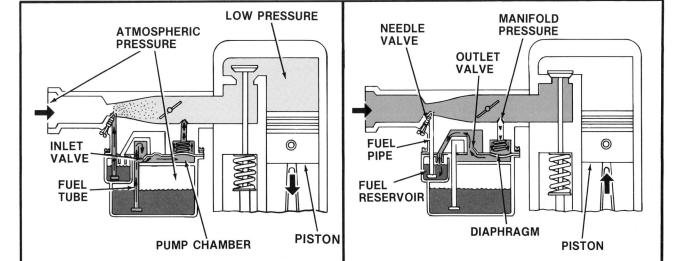

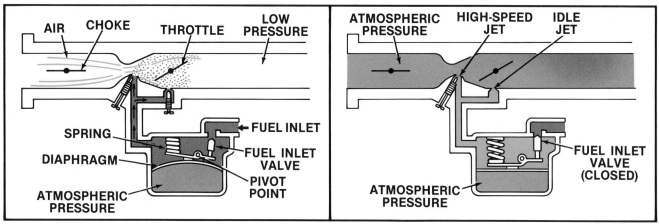

Fig. 38 — Diaphragm Carburetor Operation

This reservoir provides a constant fuel level for a second fuel pipe which is connected to the needle valve passage. The second fuel pipe and the rest of the carburetor operates exactly as described in vacuum suction-lift carburetors.

Diaphragm-Type Carburetor

The diaphragm-type carburetor is equipped with a flexible diaphragm which is exposed to crankcase pressure on one side and atmospheric pressure on the other (Fig. 38). This diaphragm and its accompanying fuel chamber take the place of the float and float bowl in a float-type carburetor.

The advantage of this type of carburetor is that it is functional in any position. An example of a diaphragm carburetor equipped with an integral fuel pump is described in the section on chain saw engines.

OPERATION

As the pressure differential between the crankcase and atmospheric pressure increases the diaphragm flexes

Fig. 39 — Check Ball in Main Jet

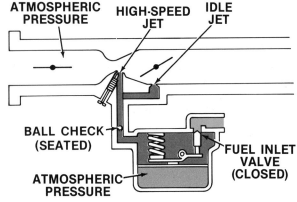

inward and lifts the inlet needle from its seat (Fig. 38, left). This permits fuel to enter the fuel chamber.

As the chamber fills and the pressure differential decreases the diaphragm moves outward, the inlet needle reseats and cuts off the fuel flow (Fig. 38, right).

The operation of the carburetor is otherwise similar to that of a float-type carburetor including the action of the one or more idle ports and the main jet. Idle and high speed mixtures are controlled by separate needle valves. A main jet ball check valve is sometimes located at the lower end of the main jet (Fig. 39). This check maintains the fuel level in the jet for rapid throttle response.

AUTOMATIC CHOKES

Automatic chokes found on compact equipment are usually **diaphragm-type chokes.** The diaphragm-type choke does not depend on engine temperature to position it. A diaphragm is located under the carburetor and a linkage from the top of the diaphragm is connected to the choke lever arm (Fig. 40).

The underside of the diaphragm faces a pocket which is connected to the engine side of the carburetor throat by a passage. A spring located in this pocket, or vacuum chamber, forces the diaphragm upward and the choke closed when the engine is not running (Fig. 40, left). When the engine is started, the pressure differential on the diaphragm causes it to move downward, compressing the spring and opening the choke (Fig. 40, right).

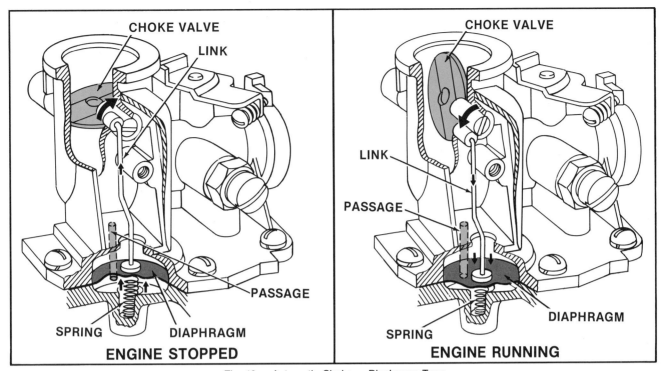

Fig. 40 — Automatic Choke — Diaphragm Type

GOVERNORS

As the throttle valve is opened, air moves faster through the carburetor throat and draws more fuel into the engine. Too much fuel entering the engine could cause it to overspeed which could result in damage to components.

Governors are used to limit the maximum speed of the engine to a specified rpm. Two types of governors are most common on compact equipment gasoline engines. They are:

• **Air vane**

• **Flyweight**

Both types are used to control engine speed and the operation will be discussed.

Air-Vane Governors

Air-vane type governors are generally found on air-cooled engines which use fins on the flywheel to generate air flow for engine cooling. The air flow generated by the rotating flywheel actuates the air-vane type governor.

The air vane is placed near the flywheel in the path of the cooling air (Fig. 41). The vane is connected through linkages to the throttle valve. As the engine speed changes (the speed of flywheel rotation), the speed of the air flow also changes. This affects the position of the air vane and thus the position of the throttle.

When the engine is stopped, a spring holds the throttle valve open (Fig. 41, left). When the engine is started, and speeds up, progressively greater air speed works against the air vane.

As the air vane moves, it begins to close the throttle from the wide open position to the position which allows the engine to run at the speed selected by the operator (Fig. 41, right).

When this speed is reached, the air flow from the flywheel against the air vane equals the governor spring tension.

Flyweight Governors

The flyweight governor assembly consists of the following components (Fig. 42):

• **Plunger and support shaft**

• **Flyweights**

• **Gear**

• **Lever**

• **Arm with spring**

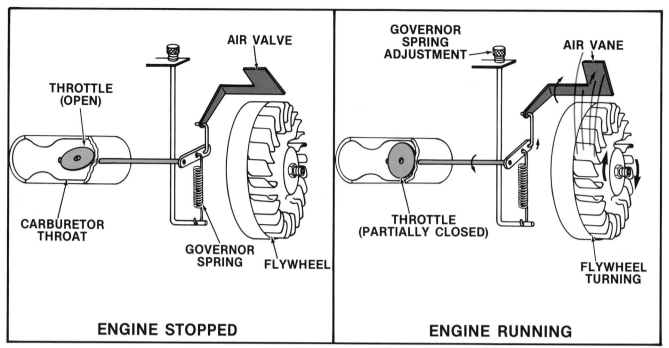

Fig. 41 — Air - Vane Governor Operation

When the engine is stopped, the governor spring tension pivots the governor arm and lever which forces the flyweights to remain close to the support shaft (Fig. 42, left). In this position, the throttle valve is in the fully open position.

When the engine starts, the camshaft and crankshaft begin to turn. The governor gear also begins to turn because it is in mesh with the camshaft gear.

As the engine picks up speed, the camshaft and governor begin to turn faster. This increase in speed

also increases the centrifugal force on the flyweights. The flyweights begin to pivot outward, and force the plunger against the governor lever (Fig. 42, right).

The governor lever then pivots the governor arm, which is linked to the throttle valve, and begins to close the throttle valve. When the engine reaches the desired speed, the centrifugal force on the flyweights equals the governor spring tension and the engine continues to operate at that speed.

Fig. 42 — Flyweight Governor Operation

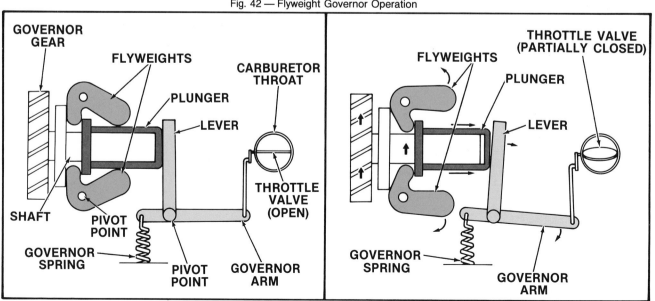

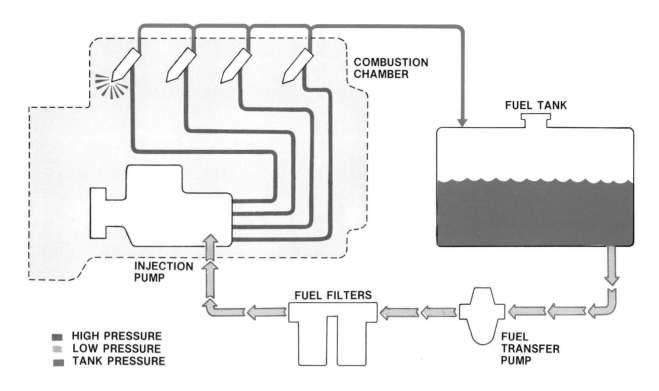

Fig. 43 — Diesel Fuel System

DIESEL FUEL SYSTEMS

Diesel fuel systems generally consist of the following components (Fig. 43):

- **Fuel tank**

- **Fuel lines**

- **Fuel filter**

- **Fuel injection pump (with transfer pump)**

- **Fuel injection nozzles**

- **Governor**

- **Turbocharger**

- **Air cleaner**

The operation of the fuel tank and air cleaner is similar to those in gasoline fuel systems and will not be covered here. All other components have characteristics unique to the diesel fuel system.

SYSTEM OPERATION

In diesel fuel system operation, fuel flows by gravity pressure from the FUEL TANK to the transfer pump.

The TRANSFER PUMP pushes the fuel through the FILTERS, where it is cleaned.

The fuel is then pumped on to the INJECTION PUMP where it is put under high pressure and delivered to each injection nozzle in turn.

The INJECTION NOZZLES atomize the fuel and spray it into the combustion chamber of each cylinder.

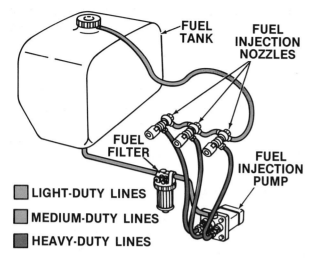

LIGHT-DUTY LINES
MEDIUM-DUTY LINES
HEAVY-DUTY LINES

Fig. 44 — Three Types of Diesel Fuel Lines

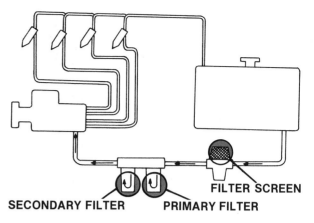

Fig. 45 — Progressive Stages of Fuel Filtration

DIESEL FUEL SYSTEM COMPONENTS

Each component of the diesel fuel system (listed earlier) has a specific function.

FUEL LINES

Fuel lines for diesel applications are of three types (Fig.44):

Light-duty lines — Used for return of leak-off fuel to tank or fuel pump. These lines carry little or no pressure.

Medium-duty lines — Carry fuel under relatively low pressure from the tank and fuel pump to the injection pump. Similar to the lines used on gasoline engines.

Heavy-duty lines — Carry high pressure (normally 1,000 - 2,000 PSI (6900 to 13 800 kPa) fuel from the injection pump to the injectors.

FUEL FILTERS

Fuel filtration is very important in diesel operation because:

• *It is more important to filter out water and contaminants to avoid damaging injection components*

• *Injection parts are precision made*

As a result, diesel fuel may be filtered not once, but several times in some systems.

A system (Fig. 45) might have three stages of progressive filters:

1. Filter screen at tank or transfer pump — removes large particles.

2. Primary filter — removes most small particles.

3. Secondary filter — removes tiny particles.

Most diesel filters have a water trap where water and heavy sediment can settle and be drained (Fig. 46).

Types Of Filters

Filtration removes suspended matter from the fuel. Some filters will also remove soluble impurities.

Fig. 46 — Diesel Fuel Filter

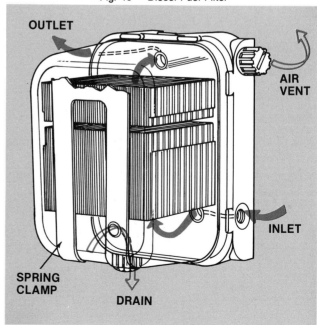

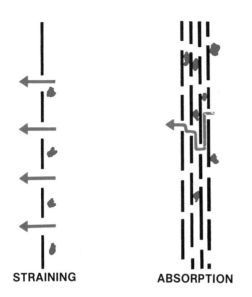

STRAINING   ABSORPTION

Fig. 47 — Methods of Fuel Filtration

Filtration can be done in three ways:

- **Straining**

- **Absorption**

- **Separation**

Straining is a mechanical way of filtering. It uses a screen which blocks and traps particles larger than the openings (Fig.47). The screen may be of wire mesh for coarse filtering or of paper or cloth for finer filtering.

Absorption is a way of trapping solid particles and some moisture by getting them to stick to the filtering material — cotton waste, cellulose, woven yarn, or felt (Fig. 47).

Separation is a method of removing water from the fuel. By treating a paper filter with chemicals, water droplets can be formed and separated when they drip into a water trap. (The filter also removes solid particles by one of the other methods of filtration.)

INJECTION PUMPS

After the fuel has been filtered it is transferred to the fuel injection pump for injection through nozzles into the combustion chambers.

Types Of Injection Pumps

The types of pump and nozzle arrangements most commonly used in compact equipment are (Fig. 48):

- **In-line type**

- **Distributor type**

The in-line injection pump uses a pump for each cylinder. These pumps are combined into one body and a single camshaft actuates the pumps. This permits the use of a single drive within the engine to make unit servicing and timing easier. A separate injection nozzle is used for each cylinder.

Fig. 48 — Pump and Nozzle Arrangements

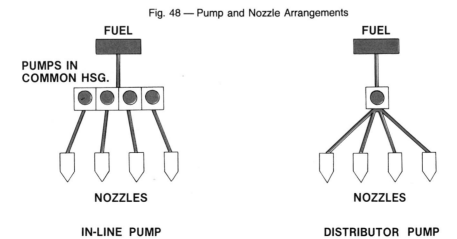

FUEL   FUEL

PUMPS IN
COMMON HSG.

NOZZLES   NOZZLES

IN-LINE PUMP   DISTRIBUTOR PUMP

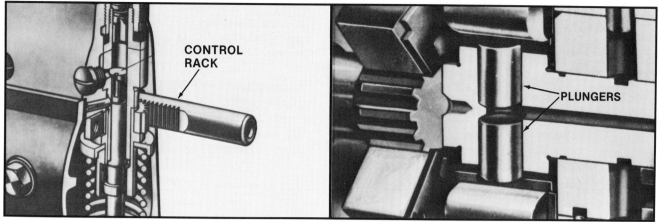

Fig. 49 — Control Rack (left) and Plungers (right)

The distributor-type injection pump uses a single pump and a distribution system to supply fuel to each cylinder in firing order. A separate injection nozzle is again used for each cylinder.

Fuel injection systems which contain different types of injection pumps may also contain different features suited to that type of pump. All fuel injection systems, however, have some common features. The following section will discuss these common features. Coverage of the operation of each type of pump will follow.

REQUIREMENTS FOR FUEL INJECTION

The basic functions of all fuel injection systems are to:

- *Supply the correct quantity of fuel to the engine*
- *Time fuel delivery and control delivery rate*
- *Pressurize fuel*
- *Break up or atomize fuel*
- *Distribute fuel evenly throughout the cylinder*
- *Govern engine speed*

Various components of the fuel injection system operate to fulfill these basic requirements. The following discussion will examine how these components work to accomplish these basic requirements.

Supply Correct Quantity Of Fuel

The fuel injection system must deliver the exact amount of fuel to each cylinder during each engine cycle. The in-line injection pump uses a **control rack** to regulate the amount of fuel delivered while the distributor pump relies on the length of a **plunger stroke** (Fig. 49).

Fuel Delivery Rate And Timing

Fuel delivered to the combustion chamber at the wrong time during the power stroke causes a loss of power. A **delivery valve** (Fig. 50) used in both types of pumps works with the control rack or plungers to obtain the correct fuel delivery rate and timing. Smooth operation from each cylinder depends on the amount of time taken to inject fuel. The higher the engine speed, the faster the fuel must be injected. This is controlled by load and speed **advance mechanisms** found in some pumps (Fig. 51).

Pressurize Fuel

Fuel is under high pressure when it is injected into the cylinder. The pumping elements in both types of pumps **(plunger and delivery valve)** work to pressurize the fuel. High pressure is required to force fuel into the combustion chamber which is pressurized by piston movement. The high pressure also aids in obtaining the desired swirl action of fuel and air in the combustion chamber for smooth combustion.

Fig. 50 — Delivery Valve

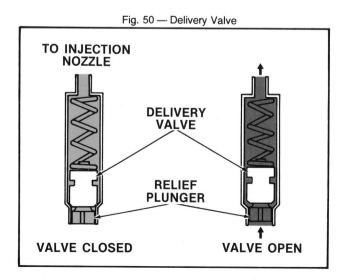

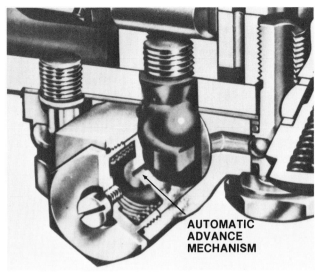

Fig. 51 — Advance Mechanism

Atomize And Distribute Fuel

Not only must the fuel be injected under high pressure into the combustion chamber, it must also be broken up into fine particles. This fuel atomization allows the fuel to mix thoroughly with the air for complete combustion (Fig. 52). Atomization also allows fuel to spread evenly in the combustion chamber. Atomization and even fuel distribution makes the engine run smoothly and develop maximum power.

Fig. 52 — Nozzles Atomize Fuel

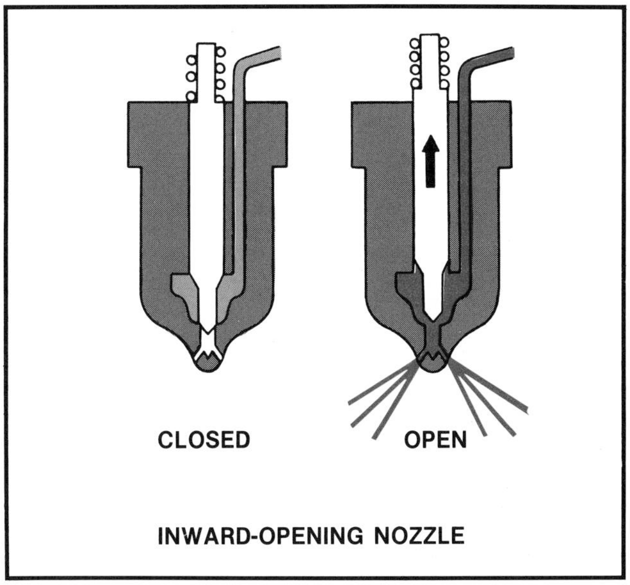

CLOSED   OPEN

INWARD-OPENING NOZZLE

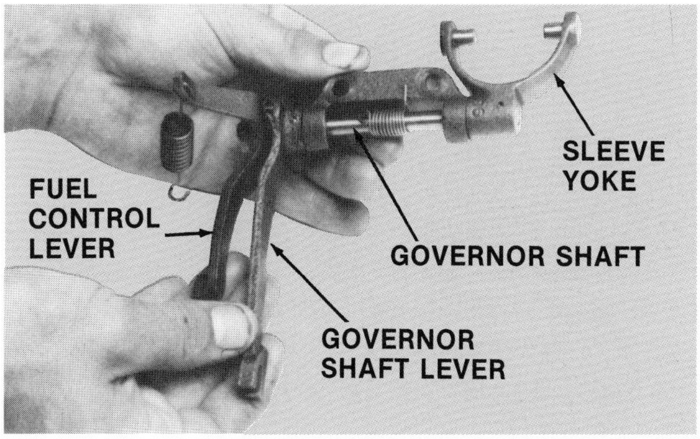

Govern Engine Speed

A **governor** (Fig. 53) maintains a nearly constant engine speed once the desired speed is set by the operator. When the speed control lever is positioned, the governor operates the injection pump control racks or metering valve to maintain a nearly constant speed. The governor varies the amount of fuel supplied to the engine to satisfy varying load demands.

IN-LINE INJECTION PUMPS

The in-line injection pump uses an individual pump for each cylinder (Fig. 54).

The individual pumping elements are usually mounted together in packs or in line as shown.

In-Line Pump Components

Main parts of the pump as shown in Fig. 54 are:

- **Pump housing**
- **Fuel transfer pump**
- **Camshaft**
- **Plunger and barrel**
- **Control rack**
- **Delivery valve**

Fig. 53 — Part of Governor Assembly

SLEEVE YOKE

FUEL CONTROL LEVER

GOVERNOR SHAFT

GOVERNOR SHAFT LEVER

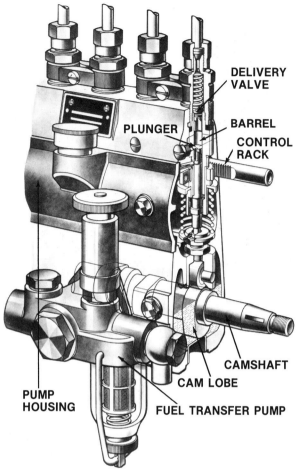

DELIVERY VALVE

PLUNGER

BARREL

CONTROL RACK

CAMSHAFT

CAM LOBE

PUMP HOUSING

FUEL TRANSFER PUMP

Fig. 54 — In-Line Injection Pump

The typical pump shown is a single-acting plunger type driven by the engine.

Fuel Flow

Fuel flows from the fuel tank to the fuel transfer pump. The transfer pump sends the fuel on at low pressure to fuel filters and then back to the injection pump.

Each pumping element then meters the fuel at high pressure to its engine cylinder.

A hand primer is located on the fuel transfer pump. This primer can be operated by hand to pump fuel when bleeding the system or when the fuel lines are disconnected.

Fuel Transfer Pump

The fuel transfer pump (Fig. 54) assures that fuel is supplied to the injection pump at all times. The fuel is pumped through the fuel filters on its way, assuring that only clean fuel reaches the injection pump.

The transfer pump is a single-acting, piston-type pump mounted on the side of the injection pump, and is driven by a cam on the injection pump camshaft. All fuel flows through a preliminary filter located in the transfer pump sediment bowl.

The hand primer is screwed into the transfer pump housing. The pump is operated by unscrewing the

Fig. 55 — Injection Pump Fuel Flow

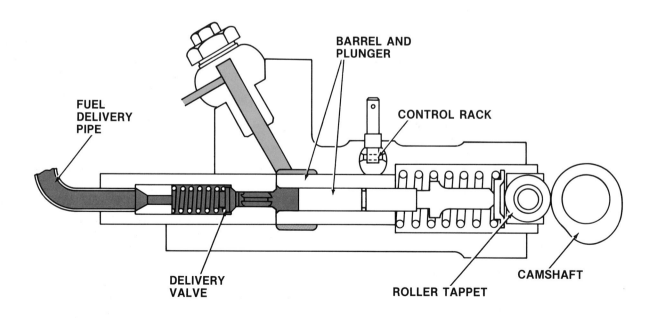

FUEL DELIVERY PIPE

BARREL AND PLUNGER

CONTROL RACK

CAMSHAFT

DELIVERY VALVE

ROLLER TAPPET

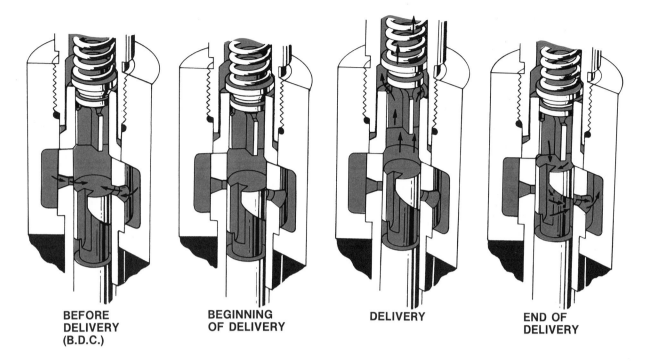

BEFORE DELIVERY (B.D.C.)   BEGINNING OF DELIVERY   DELIVERY   END OF DELIVERY

Fig. 56 — Plunger Operation (At Maximum Fuel Delivery)

knurled knob and working the plunger up and down. When the plunger is pulled up, the pump cylinder is charged with fuel. When the plunger is pushed down, this fuel is forced through the fuel filters and into the injection pump.

Injection Pump Fuel Flow

The single-acting, plunger-type injection pump has an engine-driven camshaft which rotates at one-half engine speed. Roller cam followers, riding on the camshaft lobes, operate the plungers to supply high-pressure fuel through the delivery valves to the injection nozzles (Fig. 55).

A governor-operated control rack is connected to the control sleeves and plungers, to regulate the amount of fuel delivered to the engine.

Engine lubricating oil is piped into the injection pump to provide splash lubrication to the working parts.

Pumping Elements

The pumping elements (Fig. 55) are a plunger and barrel — one set to supply each engine cylinder. The plunger is precisely fitted to the barrel by lapping to provide clearance of about 0.0001 inch (0.002 mm). With such small clearance, perfect sealing results

without special sealing rings. A pumping element is always replaced as a complete unit (barrel with matching plunger).

The plungers are operated at a constant stroke; that is, they move the same distance each time the cam actuates them.

To vary the amount of fuel delivered per stroke for satisfying varying load demands, the upper part of the plunger is provided with a vertical channel extending from its top face to an annular groove, the top edge of which is milled in the form of a helix (called the control edge). On top of the plunger is a machined notch called the retard notch, which retards the injection timing for starting the engine.

The barrel has either a single or double control port, depending on the design. An annular groove inside the barrel routes any fuel leakage between the plunger and barrel through a hole in the barrel back to the fuel gallery. The top of the barrel is closed by a spring-loaded valve called the delivery valve.

Operation Of Pumping Elements

When the plunger is at the bottom of its stroke (Fig. 56) fuel fills the space above the plunger, the vertical

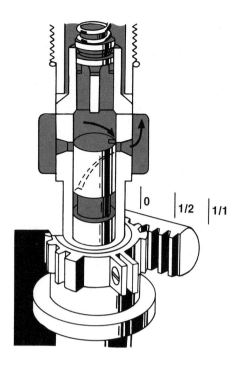

Fig. 57 — No Fuel Delivery

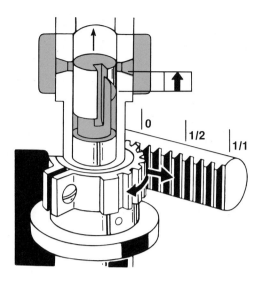

Fig. 58 — Partial Fuel Delivery

slot, and the cut-away area below the helix. The fuel flows into the area from the fuel gallery through ports in the barrel.

As the plunger moves upward, it closes the barrel ports, and discharges the fuel trapped in the pressure area through the delivery valve and line to the injection nozzle. Delivery of fuel stops as soon as the control edge of the helix uncovers the control port. Fuel then flows out through the vertical slot and annular groove back into the fuel gallery.

Plunger Positions

NO FUEL DELIVERY

No fuel delivery occurs when the plunger is rotated to a position where the vertical slot in the plunger aligns with the control port (Fig. 57). Since the vertical slot prevents the control port from being covered, pressure cannot build up. Hence, no fuel can be forced to the injection nozzles.

PARTIAL FUEL DELIVERY

Partial fuel delivery occurs at any position of the plunger between no delivery and maximum delivery depending on the position of the helix in relation to the control port (Fig. 58).

MAXIMUM FUEL DELIVERY

Maximum fuel delivery (Fig. 59) occurs when the plunger is rotated to a position where the control port is covered by the plunger as it moves upward, for the greatest possible distance (effective stroke) permitted by the control rack.

Fig. 59 — Maximum Fuel Delivery

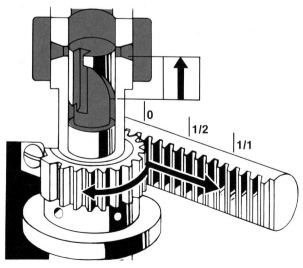

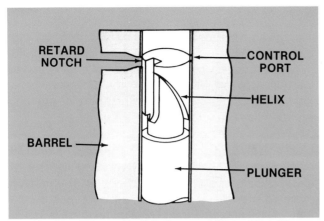

Fig. 60 — "Excess Fuel" Delivery and "Retard Notch" (Exaggerated)

RETARD NOTCH

A "retard notch" (Fig. 60) on the upper face of the plunger also aids in starting a cold engine. This retard notch is aligned with the control port when starting the engine, which means that the plunger must move farther upward before the control port will be completely closed. This delay in port closing results in the desired timing retardation.

When the "retard notch" is aligned with the control port, the plunger helix is designed to be in the "excess fuel" position. In this way, both "excess fuel" and retarded of injection timing work together when starting the engine.

Control Rack And Sleeve

The control rack (Fig. 61) is connected to the governor by linkage which moves the rack to regulate the speed of the engine.

The sleeve, which is actuated by the control rack, is fitted over the barrel and accepts the vanes on the plunger. This provides positive rotation of the plunger when the sleeve is rotated by the rack.

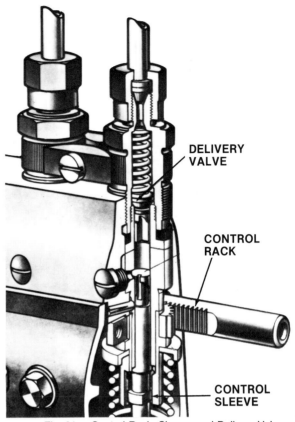

Fig. 61 — Control Rack, Sleeve, and Delivery Valve

Other pumps have a pin at the upper end of the control sleeve which engages a slot in the control rack.

Also affecting the control rack travel, is the starting fuel control shaft which permits the rack to move to the "excess fuel" position described earlier.

Fig. 62 — Delivery Valve

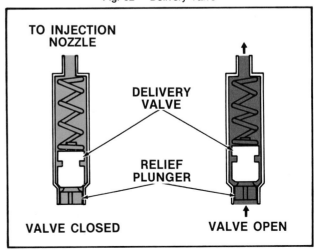

Delivery Valve

The delivery valve (Fig. 62) is guided by its stem in the valve housing.

During the fuel delivery stroke, the valve is forced off its seat and fuel is forced along the longitudinal grooves over the valve face into the delivery line.

When the helix on the pump plunger uncovers the control port, the pressure drops suddenly in the pump barrel. The pressure in the delivery line and the valve spring force the valve back on its seat. When this happens, the movement of the valve relieves the pressure in the delivery line, thus preventing dribbling at the injection nozzle.

To do this, the delivery valve is constructed with a relief plunger, which fits into the valve holder. When the delivery stroke ends, and the valve starts to resume its seated position, the relief plunger will move into the bore of the valve holder, thus sealing the delivery line from the pressure chamber. After the relief plunger has entered the bore of the valve housing, the valve seats firmly. Now the space for fuel in the delivery line is increased by an amount equal to the volume of the relief plunger. The effect of this increase in volume is a sudden pressure drop in the delivery line, causing the injection nozzle valve to close instantly.

NOTE: *For service instructions on in-line injection pumps, refer to the pump technical manual.*

Governors

The governor regulates the amount of fuel the injection pump supplies to the engine. The amount of fuel supplied depends on the speed of the engine and the load under which it operates. Many different types of governors exist, however, the principles of operation are much the same for each type.

Mechanical governors are most common on compact equipment. Pneumatic and hydraulic governors are found on farm and industrial equipment. The following discussion will cover the operation of one type of mechanical governor found on a compact utility tractor.

The principles of governor operation are generally the same even though there are many types of mechanical linkages. Read the technical manual for governor operation for a specific machine.

COMPONENTS AND OPERATION

Main components of a mechanical governor are shown in Fig. 63.

At engine start-up, the hand lever is operated so as to pull the speed control lever down (Fig. 63). This in turn will pull the fuel control lever forward causing the index pin to contact and move the governor shaft lever. This moves the injection pump rack forward to supply fuel to the engine.

Since the governor flyweights and sleeve have not moved (because engine is not running), the governor spring will further move the shaft lever and pump rack forward. This allows the engine to receive enough fuel for easier starting.

Fig. 63 — Governor Components

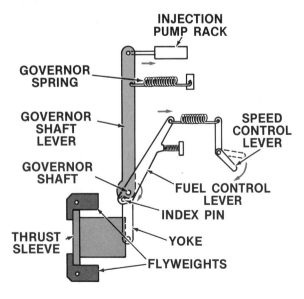

INJECTION
PUMP RACK

GOVERNOR
SPRING

GOVERNOR
SHAFT
LEVER

SPEED
CONTROL
LEVER

GOVERNOR
SHAFT

FUEL CONTROL
LEVER

INDEX PIN

THRUST
SLEEVE

YOKE

FLYWEIGHTS

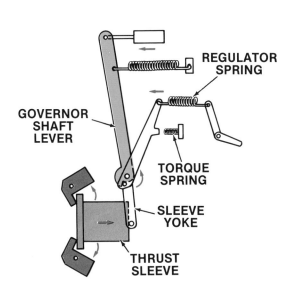

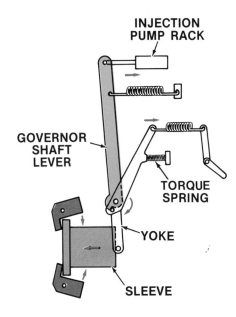

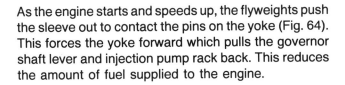

Fig. 64 — Governor Operation — Engine Startup

Fig. 65 — Governor Operation Under Load

As the engine starts and speeds up, the flyweights push the sleeve out to contact the pins on the yoke (Fig. 64). This forces the yoke forward which pulls the governor shaft lever and injection pump rack back. This reduces the amount of fuel supplied to the engine.

When the governor shaft lever is pulled back, it contacts the index pin on the fuel control lever. Since the position of the hand lever initially determines the position of the fuel control lever, the regulator spring expands. The tension on the regulator spring balances the force being applied by the governor sleeve.

The centrifugal force of the governor flyweights and regulator spring force must be balanced to obtain a constant speed. When either one of the two forces exceeds the other, a change in fuel delivery will take place as long as the demands are within operational limits (engine not overloaded).

As a load is applied to the engine, engine speed is reduced because the engine is not receiving enough fuel to overcome the load. This will reduce the centrifugal force of the governor flyweights causing the thrust sleeve and yoke to move back (Fig. 65). With the thrust sleeve moved away from yoke, there will be less force to resist the tension of the regulator spring.

The lessened force on the regulator spring will allow the fuel control lever to move forward and contact the torque spring. At the same time, the shaft lever will be moved along with the injection pump rack to deliver more fuel to the engine.

With the fuel control lever just contacting the torque spring, the engine will be running at rated speed. If more load is applied to the engine, the control lever will move further ahead compressing the torque spring. The shaft lever and injection pump rack will also move to provide more fuel. The torque spring restricts the control lever from moving too far or too quickly which would deliver too much fuel to the engine. Overfueling may result in damage to engine parts.

As the load on the engine stabilizes or decreases, the centrifugal force of the flyweights will again balance with the regulator spring tension. Less fuel will be delivered to the engine to obtain a constant engine speed.

If an overload condition exists, the injection pump will continue to deliver fuel at its maximum rate until the overload condition is removed or the tractor stalls.

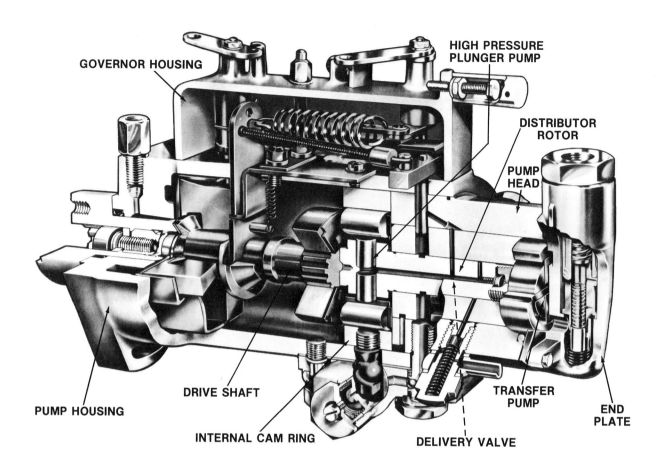

GOVERNOR HOUSING

HIGH PRESSURE
PLUNGER PUMP

DISTRIBUTOR
ROTOR

PUMP
HEAD

PUMP HOUSING

DRIVE SHAFT

INTERNAL CAM RING

TRANSFER
PUMP

END
PLATE

DELIVERY VALVE

Fig. 66 — Distributor Pump Components

DISTRIBUTOR INJECTION PUMP

The type of distributor injection pump commonly used is a primary metering pump. This means that fuel is metered at low pressure before being pressurized for injection.

Distributor Pump Components

Main components of the distributor pump are (Fig. 66):

Drive shaft — Provides engine drive to pump components.

Transfer pump — A positive displacement vane type pump which transfers fuel to the metering valve, governor valve, and high pressure pump.

Distributor rotor — Distributes fuel to the individual injector nozzles. Provides mounting and drive for the transfer pump, plunger pump, and delivery valve.

High pressure pump — A radial, plunger pump which supplies high pressure fuel to the distributor.

Delivery valve — A single delivery valve is located in the main output passage.

Internal cam ring — A ring fitted with internal cam lobes which activate the plungers on the high pressure pump. There are two opposed lobes for each engine cylinder. The cam ring is mounted in the main pump housing and can be rotated hydraulically within limits for automatic timing adjustment.

Pump head — The pump head is in the main pump housing and contains the bores for the distributor, charging ports, fuel delivery passages, the discharge outlets, and the metering valve bore.

End plate — The end plate seals the end of the pump head, and contains the fuel inlet passage, fuel strainer, transfer pump, and transfer pump regulator.

Governor — The governor is a mechanical-weight type which controls engine speed by means of a positive linkage to the metering valve.

Housing — The aluminum housing provides drive shaft support bearings and contains the governor, high pressure pump, and cam ring. The housing is filled with fuel oil at all times during operation.

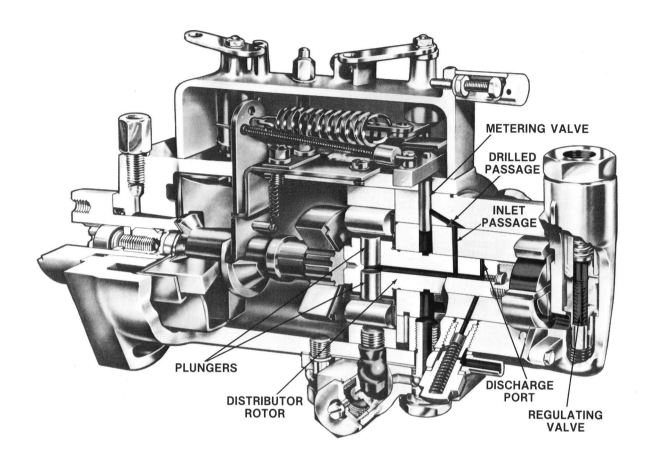

METERING VALVE

DRILLED
PASSAGE

INLET
PASSAGE

PLUNGERS

DISTRIBUTOR
ROTOR

DISCHARGE
PORT

REGULATING
VALVE

Fig. 67 — Distributor Pump Operation

Distributor Pump Operation

Fuel is drawn from the supply tank into the pump, through the inlet strainer by the vane type fuel transfer pump (Fig. 67).

Since transfer pump displacement greatly exceeds the injection requirements, a large percentage of fuel is bypassed through the regulating valve back to the inlet side. The flow of this positive displacement pump increases with speed and the regulating valve is designed so transfer pump pressure also increases with speed.

Fuel, under transfer pump pressure, is forced through drilled passages in the pump head to the metering valve.

The radial position of the metering valve, controlled by the governor, regulates the flow of fuel through inlet passages into the pumping cylinders.

As the distributor rotor rotates, the inlet passages move out of registry and the single discharge port is opened. The cam rollers contact the cam lobes forcing the plungers together. Fuel trapped between the plungers is then delivered to the nozzle.

The pump is lubricated by the fuel. As fuel at transfer pump pressure reaches the charging ports, slots on the rotor shank allow fuel and any trapped air to flow into the pump housing cavity.

In addition, an air bleed in the pump head connects the outlet side of the transfer pump with the pump housing cavity. This allows air and some fuel to bleed back to the fuel tank through a return line. The bypassed fuel helps lubricate the internal parts.

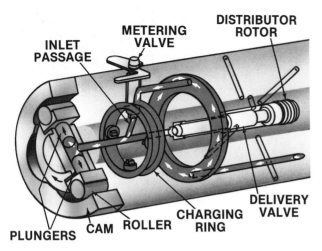

Fig. 68 — Charging Cycle

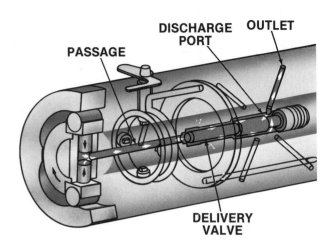

Fig. 69 — Discharging Cycle

Charging And Discharging

Figs. 68 and 69 show the fuel flow during the charging and discharging cycles.

CHARGING CYCLE

As the distributor rotor revolves (Fig. 68) the inlet passage in the rotor registers with a charging port of the charging ring. Fuel under pressure from the transfer pump and controlled by the opening of the metering valve flows into the pumping cylinders, forcing all plungers apart.

The plungers move outward a distance proportionate to the amount of fuel required for injection on the following stroke. If only a small quantity of fuel is admitted into the pumping cylinders, as at idling, the plungers move out very little. Maximum plunger travel and, consequently, maximum fuel delivery is limited by adjusting leaf springs which contact the edge of the roller shoes. Only when the engine is operating at full load will the plungers move to the most outward position.

Note in Fig. 68 that the inlet passage in the rotor is in registry with a port in the charging ring, but the rotor discharge port is not in registry with a head outlet. Note also that the rollers are between the cam lobes. Compare their relative positions in Figs. 68 and 69.

DISCHARGE CYCLE

As the rotor continues to revolve (Fig. 69), the inlet passage moves out of registry with the charging port.

For a brief interval the fuel is trapped until the rotor discharge passage registers with one of the head outlets.

As this registration takes place, both sets of rollers contact the cam lobes and are forced together.

During this stroke the fuel trapped between the plungers is forced through the axial passage of the rotor and flows through the rotor discharge passage to the injection line.

Delivery to the line continues until the rollers pass the highest point on the cam lobe and are allowed to move outward.

The pressure in the axial passage is then relieved, allowing the injection nozzle to close.

Delivery Valve

The major function of the delivery valve is to rapidly decrease the injection line pressure after injection to a pressure lower than that of the nozzle closing pressure.

This reduction in pressure causes the nozzle valve to return rapidly to its seat, achieving sharp delivery cut-off and preventing dribble of fuel into the engine combustion chamber.

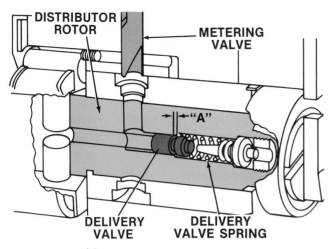

Fig. 70 — Delivery Valve Operation

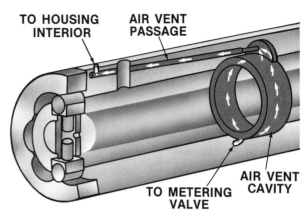

Fig. 71 — Return Oil Circuit

The delivery valve (Fig. 70) is located in the center of the distributor rotor. It requires no seat — only a shoulder to limit travel.

Since the same delivery valve performs the function of reducing the pressure for each injection line, the pressure reduction will not vary from cylinder to cylinder. This results in a smooth-running engine at all loads and speeds.

When injection starts, fuel pressure moves the delivery valve slightly out of its bore and adds the volume of its displacement (section "A") to the enlarged cavity of the rotor occupied by the delivery valve spring. This displaces a similar volume of fuel in the spring cavity before delivery through the valve ports starts.

At the end of injection, the pressure on the plunger side of the delivery valve is quickly reduced because the cam rollers move to the lower position on the cam lobes. The plungers move apart. As they move apart, some fuel from the discharge port is drawn back toward the plungers. This quickly reduces pressure in the delivery valve. The delivery valve spring then moves the valve back to the shoulder.

As the valve returns to its closed position, its displacement (section "A") is removed from the spring cavity. Since the rotor discharge port is still partly in registry, some fuel rushes back out of the injection line to fill the volume left by the retreating delivery valve.

After this, the rotor ports close completely and the remaining fuel is trapped in the injection line.

Return Oil Circuit

Fuel under transfer pump pressure is discharged into a cavity in the pump head (Fig. 71).

The upper half of this cavity connects with a vent passage. Its volume is restricted by a wire to prevent undue pressure loss.

The vent passage is located behind the metering valve bore and connects with a short vertical passage entering the governor linkage compartment.

Should air enter the transfer pump because of suction-side leaks, it immediately passes to the air vent cavity and then to the vent passage as shown.

Air and a small quantity of fuel then flow from the housing to the fuel tank through the return line.

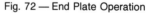

Fig. 72 — End Plate Operation

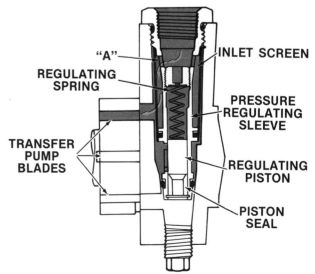

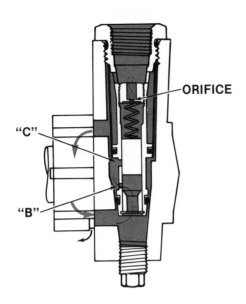

Fig. 73 — Pressure Regulating Valve Operation

End Plate Operation

The end plate (Fig. 72) has three basic functions:

1. *Provides fuel inlet passages and houses pressure regulating valve*

2. *Covers the transfer pump*

3. *Absorbs end thrust of drive and governor*

During hand priming (Fig. 69), the fuel flows into the inlet side of the transfer pump through the port "A." Priming of the rest of the system is accomplished when the pump is rotated by the engine starting motor or on the test bench.

Fig. 73 shows the operation of the pressure regulating valve while the pump is running.

Fuel pressure from the discharge side of the transfer pump forces the piston up the sleeve against the regulating spring. As pressure increases, the regulating spring is compressed slightly until the lower edge of the

regulating piston starts to uncover port "B," in Fig. 73. Since the pressure on the piston is opposed by the regulating spring, the delivery pressure of the transfer pump is controlled by the spring rate and size and number of regulating ports.

A high-pressure relief port "C" in the sleeve, above the regulating port, prevents high transfer pump pressures if the engine or pump is accidentally overspeeded.

Viscosity Compensation

The distributor pump must work equally well with different fuels and in varying temperatures which affect fuel viscosity. A feature of the pressure regulating device offsets pressure changes caused by these viscosity differences. A thin plate containing a sharp-edged orifice (Fig. 73) is located in the bottom of the spring adjusting plug. This orifice allows fuel leakage by the piston to return to the inlet side of the pump.

Flow through such a short orifice is virtually unaffected by viscosity changes. For this reason, the fuel pressure on the spring side of the piston will vary with viscosity changes. The pressure exerted on top of the piston is determined by the flow admitted past designed clearance of the piston in the sleeve.

With cold (or heavy) fuels, very little leakage occurs past the piston and flow through the adjusting plug is a function of the orifice size. Downward pressure on the piston is thus very little.

With warm (or light) fuels, leakage past the piston increases. Pressure in the spring cavity increases also, since flow through the short orifice remains the same as with cold fuel. Thus, downward pressure, assisting the regulating spring, positions the piston so less regulating port area is uncovered below it. Pressure is thus controlled and may actually overcompensate to offset other leakages in the pump, which increase with the thinner fuels.

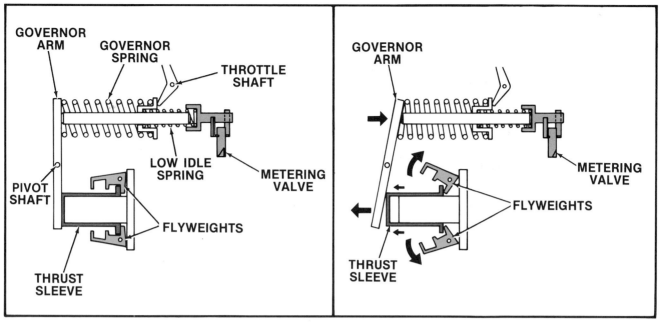

Fig. 74 — Centrifugal Governor

Centrifugal Governor

In the centrifugal governor (Fig. 74), the movement of the flyweights against the governor thrust sleeve rotates the metering valve.

This rotation varies the position of the metering valve opening with the passage from the transfer pump. This controls fuel flow to the engine.

This type of governor derives its energy from flyweights pivoting on their outer edge in the retainer. Centrifugal force tips the flyweights outward, moving the governor thrust sleeve against the governor arm. This pivots on the knife edge of the pivot shaft and, through a simple positive linkage, rotates the metering valve.

The force on the governor arm caused by the flyweights is balanced by the compression type governor spring, which is manually controlled by the speed control linkage in regulating engine speed.

A low idle spring is provided for more sensitive regulation at the low speed range.

The limits of speed control travel are set by adjusting screws for proper low idle and high idle positions.

A light tension spring takes up any slack in the linkage joints and also allows the stopping mechanism to close the metering valve without overcoming the governor spring force. Only a very light force is required to rotate the metering valve to the closed position.

Automatic Load Advance

Pumps equipped with automatic load advance (Fig. 75) permit the use of a simple hydraulic servo-mechanism, powered by oil pressure from the transfer pump, to advance injection timing.

Transfer pump pressure operates the advance piston against spring pressure as required along a predetermined timing curve.

The purpose of the load advance device is to advance injection timing as engine load decreases. This offsets the normal timing retardation that occurs at light loads.

Fig. 75 — Automatic Advance Mechanism

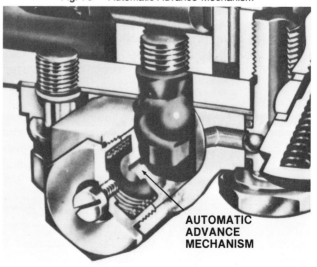

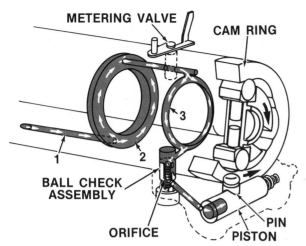

Fig. 76 — Automatic Advance Fuel Circuit

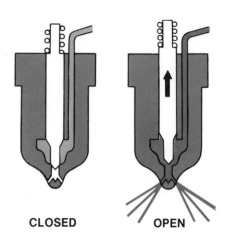

INWARD-OPENING NOZZLE

Fig. 77 — Injection Nozzle

Cam movement is induced by pressure developed at the transfer pump, admitted by the metering valve. Since the governor controls the metering valve position at all loads, it also controls the amount of fuel, under pressure, fed to the automatic advance device.

Automatic Speed Advance

Pumps equipped with automatic speed advance permit the use of a simple hydraulic servo-mechanism, powered by oil pressure from the transfer pump, to advance injection timing.

Transfer pump pressure operates the advance piston against spring pressure, as required, along a predetermined timing curve.

The purpose of the speed advance device, which responds to speed changes, is to advance the timing for the best engine power and combustion throughout the speed range.

Controlled movement of the cam in the pump housing (Fig. 75) is reduced and limited by the action of the piston of the automatic advance against a cam advance pin.

During cranking, the cam is in the retard position, since the force exerted by the advance spring is greater than that of transfer pump pressure. As the engine speed and transfer pump pressure increase, fuel under transfer pump pressure entering the advance housing behind the power piston moves the cam.

The amount of advance is limited by the length of the advance pistons. A ball check valve is provided to offset the normal tendency of the cam to return to the retard position during injection.

AUTOMATIC ADVANCE FUEL CIRCUIT

Fuel, under transfer pump pressure, is forced through the drilled passage (1, Fig. 76) located in the pump head to the annular ring (2).

Fuel then flows around and to the top of the annular ring, where it registers with the bore leading to the metering valve. The metering valve is designed to allow a quantity of fuel to flow into a second annular ring (3) which registers at the bottom with the bore of the advance clamp screw assembly.

As transfer pump pressure increases, the ball check (located in the clamp screw assembly) is lifted off its seat, allowing fuel to pass through the clamp screw into the passage located behind the power piston.

INJECTION NOZZLES

Injection nozzles (sometimes called injectors) found on compact equipment diesel engines are generally inward opening nozzles (Fig. 77). Fuel pressure acting on the lower end of the injector valve forces the valve inward against the pressure spring. As the valve is forced inward the orifices are cleared and fuel is sprayed into the combustion chamber.

The injector valve is controlled by a pressure spring which is responsible for closing the valve and holding the valve closed against compression pressures.

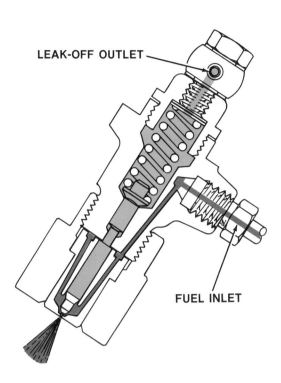

Fig. 78 — Injection Nozzle Operation

Operation

Metered fuel at high pressure from the pump enters the fuel inlet (Fig. 78).

This fuel surrounds the nozzle valve and forces the valve from its seat at a preset pressure. A measured amount of fuel then sprays out the tip into the engine combustion chamber at high velocity.

As the fuel is injected, pressure drops and the spring is able to close the valve very rapidly.

A small amount of fuel leaks past the nozzle valve and lubricates the working parts.

The excess lubricating fuel is removed from the top of the nozzle at the fuel leak-off and returns to the tank.

TURBOCHARGERS

A turbocharger is an exhaust-driven turbine which drives a compressor wheel. Turbochargers are not found on all compact equipment diesel engines but may be found on some engines.

The compressor is usually located between the air cleaner and the engine intake manifold, while the turbine is located between the exhaust manifold and the muffler.

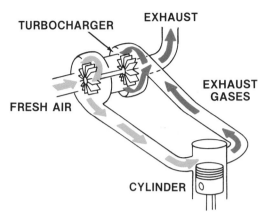

Fig. 79 — Exhaust Gases Drive Turbocharger

The prime job of the turbocharger is, by compressing the air, to force more air into the engine cylinders. This allows the engine to efficiently burn more fuel, thereby producing more power.

All of the engine exhaust gases pass through the turbine housing (Fig. 79). The expansion of these gases, acting on the turbine wheel, causes it to turn. After passing through the turbine, the exhaust gases are routed to the atmosphere. In many cases, the turbine muffles the exhaust sound, so no muffler is needed.

The compressor is directly connected to the turbine by a shaft (Fig. 80). The only power loss from the turbine to the compressor is the slight friction of the journal bearings.

Air is drawn in through a filtered air intake system, compressed by the wheel, and discharged into the engine intake manifold.

Fig. 80 — Compressor Connected to Turbine

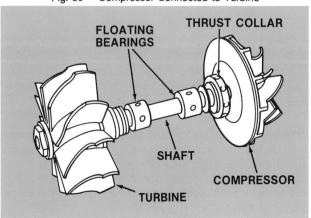

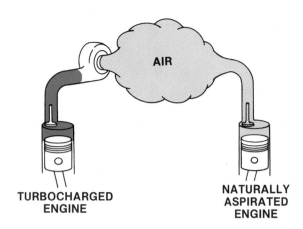

TURBOCHARGED
ENGINE

NATURALLY
ASPIRATED
ENGINE

Fig. 81 — Air Intake Comparison

The extra air provided by the turbocharger allows more fuel to be burned, which increases power output. Lack of air is one factor limiting the engine power of naturally-aspirated engines (Fig. 81).

SUMMARY

The two types of fuel systems commonly found on compact equipment are:

- **Gasoline**
- **Diesel**

GASOLINE FUEL SYSTEM

Basic components of the gasoline fuel system are:

- **Fuel tank**
- **Fuel lines**
- **Fuel pump**
- **Fuel filter**
- **Air cleaner**
- **Carburetor**
- **Governor**

A basic force-feed gasoline fuel system contains a fuel pump which draws gasoline from the fuel tank. The fuel is pumped to the carburetor where it mixes with air. The fuel-air mixture then enters the combustion chamber and is ignited.

Fuel gauges may be electric or float-type. Fuel pumps may be either electric or diaphragm type. Air cleaners may be dry-element type or oiled-element type.

Several types of carburetors are found on compact equipment gasoline engines. They are:

- **Float type — Adjustable and nonadjustable**
- **Suction-lift type — Vacuum- and pulsating-lift types**
- **Diaphragm type**

Basic part of all carburetors are:

- **Throat**
- **Venturi**
- **Throttle valve**
- **Bowl or chamber**
- **Jet**
- **Choke**

Automatic chokes are usually the **diaphragm type**. Governors may be either **air vane** or **flyweight** types.

DIESEL FUEL SYSTEM

Basic components of the diesel fuel system are:

- **Fuel tank**
- **Fuel lines**
- **Fuel filter**
- **Fuel injection pump**
- **Fuel injection nozzle**
- **Governor**
- **Air cleaner**
- **Turbocharger**

The fuel transfer pump moves fuel from the fuel tank, through the fuel filter, and to the fuel injection pump. The injection pump puts the fuel under high pressure and delivers it to each injection nozzle. The nozzle sprays the atomized fuel into the cylinder for combustion.

Basic requirements for fuel injection are to:

- *Supply the correct quantity of fuel to the engine*
- *Time fuel delivery and control delivery rate*
- *Pressurize fuel*
- *Break up or atomize fuel*
- *Distribute fuel evenly throughout the cylinder*
- *Govern engine speed*

Two types of fuel injection pumps commonly found on compact equipment are:

- **In-line**

- **Distributor**

The **turbocharger** compresses intake air, to be mixed with fuel for combustion, to force more air into the engine cylinders. It allows the engine to efficiently burn more fuel, thereby producing more power.

CHAPTER 4 REVIEW

1. Name two basic methods of transferring fuel from the fuel tank to the engine.

2. (True or false) Fuel pumps are not required in force-feed fuel systems.

3. (Fill in the blanks) Two types of air cleaners commonly found on compact equipment engines are _____ and _____.

4. Match the part of a carburetor with its definition.

Parts	Definition
1. Jet	A. A cavity where fuel is stored
2. Choke	B. A constriction in carburetor throat
3. Bowl	C. An orifice through which fuel passes.
4. Throttle valve	D. A disk or plate used to restrict air flow during cold start of engine.
5. Venturi	E. A device to control air and fuel flow after air passes the venturi

5. (True or false) A diaphragm-type choke operates by pressure differential between the carburetor throat and the diaphragm housing.

6. Name the two common types of governors found on compact equipment gasoline engines.

7. Diesel fuel systems usually contain three types of fuel lines. Give the function of each type:
    Light-duty lines — _____
    Medium-duty lines — _____
    Heavy-duty lines — _____

8. (Multiple choice) A fuel filtration method that uses a screen to block and trap particles larger than the screen openings is called:
    A. Straining
    B. Absorption
    C. Separation

9. What are the basic differences between an in-line injection pump and a distributor pump?

10. (Multiple choice) The pumping elements of an in-line injection pump are made up of:

    A. Control rack and sleeve
    B. Delivery valves
    C. Plunger and barrel
    D. Distributor rotor

11. (Fill in the blank) The control rack is connected to the _____ by a linkage which moves the rack to regulate the speed of the engine.

12. (True or false) The most common type of governor found on compact equipment diesel engines are mechanical governors.

13. (True or false) In a distributor-type injection pump, the plunger stroke is the same regardless of the amount of fuel required by the engine.

14. Explain the basic functions of the end plate of a distributor-type injection pump.

15. (Fill in the blanks) Injection nozzles most commonly found on compact equipment diesel engines are _____ _____ types.

16. (True or false) Exhaust gases drive the compressor wheel of a turbocharger.

PART 2

SERVICE, REPAIR, ADJUSTMENT, AND DIAGNOSIS OF 2- AND 4-CYCLE ENGINES

Part 2 covers engine disassembly, parts inspection, and other service procedures. Basic engine adjustments are also covered. The chapters in Part 2 are:

Chapter 5 — Servicing The 4-Cycle Engine

Chapter 6 — Servicing The 2-Cycle Engine

Chapter 7 — Servicing The Gasoline and Diesel Fuel System

Chapter 8 — Servicing The Cooling System

Chapter 9 — Servicing The Lubrication System

Chapter 10 — Diagnosis And Troubleshooting

Chapter 10 will examine some basic engine problems and give some basic guidelines on how to identify the cause of the problem. Troubleshooting charts will help to systematically diagnose these engine problems.

CHAPTER 5

SERVICING THE 4-CYCLE ENGINE

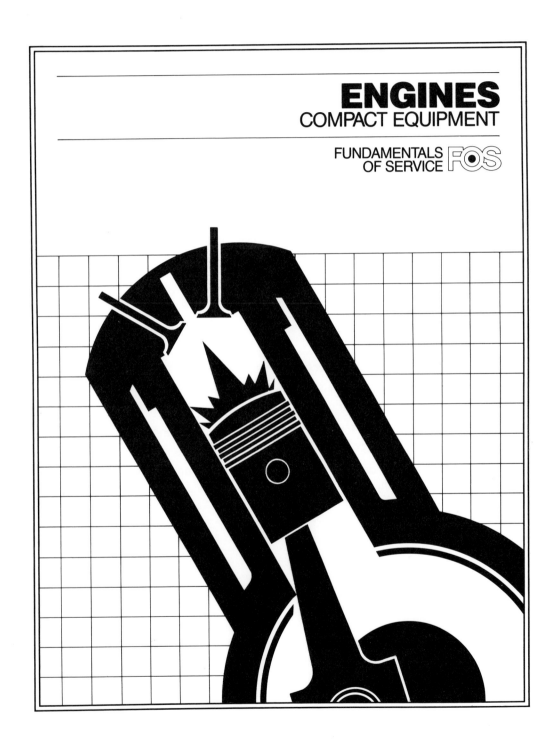

ENGINES
COMPACT EQUIPMENT

FUNDAMENTALS
OF SERVICE FOS

SKILLS AND KNOWLEDGE

This chapter contains basic information that will help you gain the necessary subject knowledge required of a service technician. With application of this knowledge and hands-on practice, you should learn the following:

• How to remove, clean, inspect and install the cylinder head

• The procedure for disassembling the valve train and inspecting for failed valves

• How to recondition valves, valve guides, and valve seats

• How to remove and install the camshaft

• The methods of adjusting valve clearance

• How to remove, clean, and inspect the cylinder block and cylinder

• The methods of reconditioning the cylinder

• How to disassemble, clean, and assemble the piston and connecting rod

• What to look for when inspecting the piston, connecting rod, and crankpin bearings for damage

• How to disassemble, inspect, and assemble the crankshaft, bearings, and flywheel

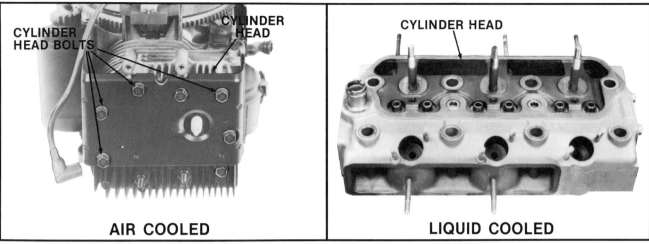

Fig. 1 — Cylinder Heads — Single Cylinder Engine (left), Multiple Cylinder Engine (right)

When service or repair is required, each engine component requires special treatment. The following pages describe disassembly, cleaning, inspection, and reassembly of each component. Engine diagnosis and troubleshooting is covered in Chapter 10.

Service procedures for many variations of 4-cycle engines will be covered in this chapter. Some of the variations include:

• **Gasoline**

• **Diesel**

• **Single cylinder**

• **Multiple cylinder**

• **Liquid cooled**

• **Air cooled**

Many of the service procedures are the same for all variations of engines. However, when a different engine style requires a different disassembly or service procedure, the text and illustrations will identify the respective style.

CYLINDER HEAD

DISASSEMBLY

First, drain the coolant from liquid cooled engines. Loosen the cylinder head bolts and disassemble the cylinder head from the cylinder block (Fig. 1). Inspect the cylinder head gasket prior to and during disassembly for signs of oil or coolant leaks or burning. Never pry on the mating surfaces of head and cylinder block to loosen a stuck cylinder head. Instead, tap the head lightly with a soft hammer.

CLEANING

Scrape all traces of old gasket material from mating surfaces of the cylinder head and the cylinder block (Fig. 2). Remove carbon deposits from the combustion chambers. To avoid gouging, especially with aluminum cylinder heads and cylinder block, use scrapers that are made of a material that will not cause damage.

Solvents are also available for removing gasket material. Make sure the proper solvent is used and make sure the solvent will not damage the cylinder head or cylinder block. Follow the engine manufacturer's recommendations for use of solvents.

INSPECTION

Check the cylinder head for cracks or other damage. Use a magnetic particle test or fluorescent dye to be sure. A severely damaged cylinder head can sometimes be repaired by a qualified specialty shop but will normally require replacement.

Fig. 2 — Scrape Gasket Material from Cylinder Head

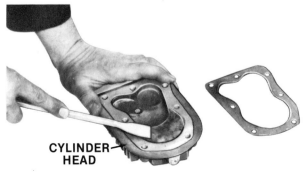

CYLINDER HEAD

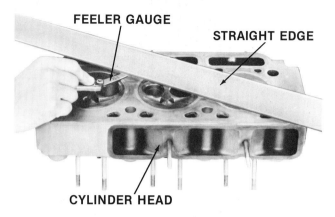

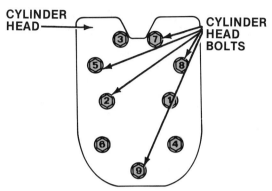

Fig. 5 — Head Bolt Tightening Sequence

Fig. 3 — Check Cylinder Head for Warping (Multiple-Cylinder Engine)

Check the cylinder head for warping. Check the cylinder heads on multi-cylinder engines with a stiff steel straightedge and a feeler gauge (Fig. 3). Place the straightedge lengthwise and side to side across the machined surfaces of the cylinder head and try to slide a thin feeler gauge under it at all points.

For cylinder heads that attach to individual cylinders, hold the cylinder head down on a flat surface and use the feeler gauge to determine flatness (Fig. 4).

REASSEMBLY

Always use a new cylinder head gasket and install it properly. Some head gaskets are easily reversed; the top of the cylinder head gasket is usually marked "TOP" to avoid errors. Follow the manufacturer's recommendations on sealant to use with the cylinder head gasket.

The technical manual illustrates the proper sequence for tightening cylinder head bolts (Fig. 5) progressively until all bolts reach the torque given in the manual.

Some manufacturers suggest that after a short run-in period the cylinder head bolts be retightened.

VALVE TRAIN

DISASSEMBLY

Disassemble the valve train according to procedures in the technical manual. Spring compressors to compress the valve spring and release tension on the keepers must fit properly (Fig. 6).

Keep the valves and valve train components in the order in which they were disassembled by making a

Fig. 4 — Check Cylinder Head for Flatness (Single-Cylinder Engine)

Fig. 6 — Valve Spring Compressor

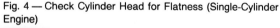

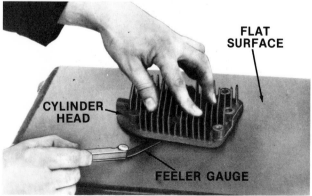

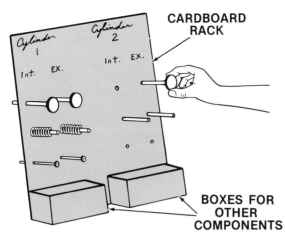

Fig. 7 — Organization Rack for Valve Train Components

Fig. 9 — Deposits Have Broken Off Valve Face

rack such as holes punched in a piece of cardboard (Fig. 7). As each component (valve, push rod, etc.) is removed from the engine, place it in its corresponding hole. Also keep cam followers and other components which don't fit the cardboard in proper order.

INSPECTION

Inspect the valve and their seats prior to cleaning. Clues may be found to indicate why the engine failed.

Check first for seat distortion. Seat distortion can be caused by:

• *Cooling system failure — heat build up causes the seats to distort or, in the case of seat inserts, loosen.*

• *Warped mating surfaces on cylinder heads or blocks mechanically distort the seats when the cylinder head bolts are tightened.*

• *Valve seats not concentric with the valve guide.*

Inspect the valve stem for deposits (Fig. 8). Excess deposits usually indicate exhaust gas leakage into the intake port area. This indicates that the valve is not seating properly.

Inspect the valve face for deposits or signs of where deposits that were once on the valve have broken off (Fig. 9). These may be caused by weak springs, too little tappet clearance, or valves sticking in the guides (sometimes due to stem deposits).

Insufficient valve clearance also holds the valve off its seat causing blowby and a burned valve (Fig. 10).

Preignition is another cause of valve burning (Fig. 11). Preignition is more a problem with gasoline engines. This burn is usually in a localized area and caused by the concentrated excessive temperatures and pressures due to detonation or explosion of the fuel subsequent to normal ignition. This problem can also damage cylinders, piston rings, and burn holes in the tops of pistons. Preignition can be caused by:

• *Improper timing*

• *Combustion chamber deposits which retain heat and ignite the fuel-air mixture prematurely.*

Fig. 8 — Deposits on Valve Stem

Fig. 10 — Valve Burned from too Little Valve Clearance

Fig. 11 — Valve Burned by Preignition

Fig. 13 — Valve Cracked by Heat

Impact break is another type of valve breakage caused by the valve striking the seat with too much impact (Fig. 14, bottom). The usual cause is too much valve clearance. Another possible cause is that the wrong spring, which had too much closing force, was installed during an earlier engine assembly.

- *Improper or damaged spark plug (gasoline engine)*

- *Excessive compression ratios. When cylinder heads or cylinder blocks are resurfaced the compression ratio can be raised sufficiently to cause a gasoline engine to detonate.*

Valve erosion is another common valve problem (Fig. 12). Erosion is generally caused by the valve overheating. This may be caused by overheated exhaust gases moving past the valve stem under the head of the valve. The gases may be overheated because of too lean of a fuel-air mixture. These gases erode the metal and can eventually cause the valve head to break off the stem.

Heat cracking of the valve (Fig. 13) or a fatigue break of the valve stem (Fig. 14, top) may be caused by worn guides or preignition.

Fig. 14 — Broken Valves — Two Causes

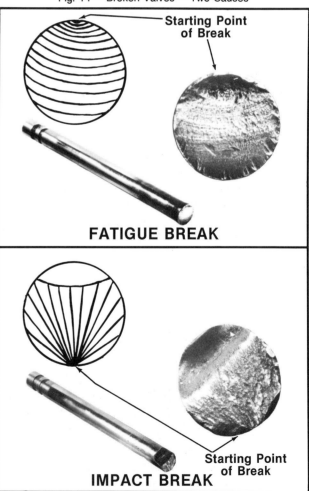

FATIGUE BREAK

IMPACT BREAK

Fig. 12 — Erosion Under Head of Valve

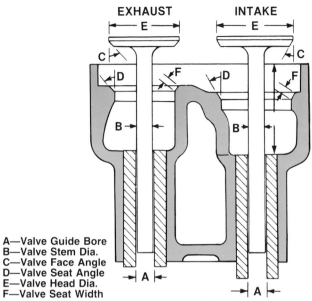

EXHAUST   INTAKE

A—Valve Guide Bore
B—Valve Stem Dia.
C—Valve Face Angle
D—Valve Seat Angle
E—Valve Head Dia.
F—Valve Seat Width

Fig. 15 — Valve Dimensions Required for Servicing

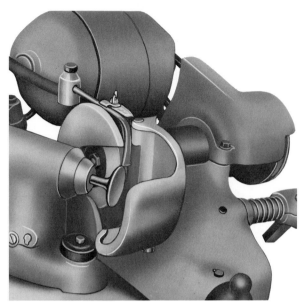

Fig. 17 — Refacing Valve

Worn valve guides can be identified by carbon deposits and wear lines in the valve guide area of the stem. Worn valve guides can be caused by dirt, worn rocker arms, poor lubrication, carbon deposits, or misaligned valve springs.

Reconditioning Valves and Guides

Before reconditioning valves and guides, check the dimensions to see if they meet the specifications in the technical manual (Fig. 15). Any part that is not within its specification must be reconditioned or replaced.

Valves are cleaned by wire brushing or a wire wheel mounted on a bench grinder. After removing deposits, polish the stem with steel wool or crocus cloth to remove scratches left by the brush. Check the valve for cracks or burns which will prevent reuse. Measure

the valve stem with a micrometer to determine if it is worn excessively (Fig. 16). Measure the stem at several points on the stem where it moves in the valve guide.

Make a final inspection of the valve while refacing. Set the angle of the refacer head as specified, or 1/2 to 1-1/2 degree greater if an interference fit is desired. Be sure the valve stem and the chuck of the facing tool (Fig. 17) are absolutely clean. Make the first grinding pass very light. If, after the first pass, less than one half of the circumference of the face has been ground check the chuck and stem for dirt. If they are clean the valve is warped or distorted and should be replaced.

Continue grinding until the face is ground to the width specified in the manual. Be sure the margin is uniform around the circumference of the valve (Fig. 18). A knife edge margin or one of uneven dimensions indicates excessive wear or warping. Replace the valve.

Fig. 16 — Measuring Valve Stem

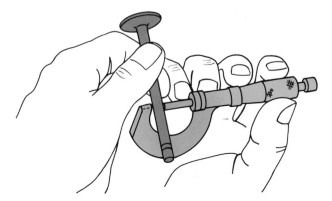

Fig. 18 — Valve Head Margin

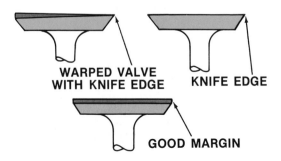

WARPED VALVE
WITH KNIFE EDGE   KNIFE EDGE

GOOD MARGIN

Fig. 19 — Cleaning Valve Guides

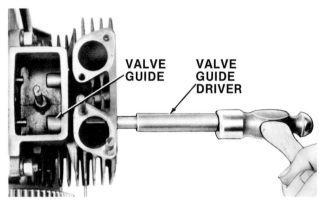

Fig. 21 — Removing Valve Guide with Driver

Reface the valve stem tip on a refacer attachment if pitting or wear is present. Use a very light cut.

Clean valve guides with a special wire brush, matched to the diameter of the guide, driven by an electric drill (Fig. 19). Dip the brush in light oil or kerosene and run through the guide until all deposits have been removed.

After cleaning, measure the valve guide with a telescoping gauge (Fig. 20). Check near both ends and at the center and measure each area twice, turning the gauge 90 degrees for the second measurement. The readings will indicate if the guide is bell-mouthed toward the ends or worn out of round.

If the guides are bell-mouthed or the tolerances between guide dimension and valve stem dimensions are greater than the maximums given in the manual, replace the guide.

Remove guides with a special puller, a piloted pressing mandrel on a press, or a valve guide driver (Fig. 21). Driving the guides out with a drift punch will damage the guide bores. Loose guides require an oversized guide or knurling the guide bores prior to reassembly.

Clean the guide bores thoroughly before installing new guides and seats. Press the new guides in making sure they are properly seated. Valve guides are often slightly compressed during insertion. Use a piloted reamer and then a finishing reamer to resize the bore to the correct dimension (Fig. 22).

Fig. 20 — Measuring Valve Guides

Fig. 22 — Piloted and Finishing Reamer

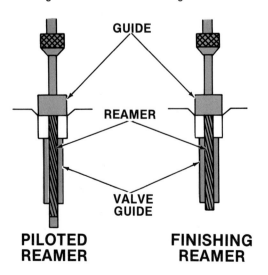

GUIDE

REAMER

VALVE GUIDE

PILOTED REAMER   FINISHING REAMER

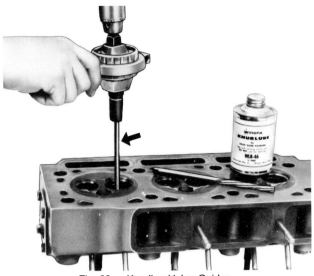

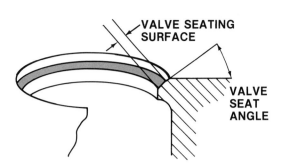

Fig. 25 — Check Valve Seat Dimensions

Fig. 23 — Knurling Valve Guides

Integral guides cannot be removed. If the guide must be reconditioned, the bore must be reamed oversized and knurled with a special knurling tool (Fig. 23). The guide could also be reamed oversized and fitted with new valves with oversized stems. Neither method is totally satisfactory since it is difficult to maintain concentricity with the valve seat. The exception would be a production type piloted reamer. It is often best to replace the short block on a compact equipment engine with this problem.

Reconditioning Valve Seats

After the guides are reconditioned, the valve seat may require reconditioning. This procedure is often not necessary unless the valve seat is damaged or excessively worn. In many cases, valve lapping, which will be discussed later in this section, is all that is required to obtain proper valve seating.

If valve seat reconditioning is required, mount the seat grinding stone or cutter (Fig. 24) on a pilot which is inserted in the bore of the valve guide to maintain alignment between guide and seat.

First use a wire brush in an electric drill to clean the entire seat area. Blow away all particles and assure that both seat area and guide are clean.

To grind or cut the valve seat, insert the cutting tool into the valve guide and assemble the proper size and angle cutter on the tool. Make a light cut and observe the results (Fig. 25). If the cut area is rough or uneven and the seat is an insert, replace the insert (Fig. 26). If the engine is not equipped with inserts, the seat must be ground deep enough to give a smooth surface of the proper width. The seat area can be counterbored and fitted with an insert but this requires special equipment and is normally only done on larger engines.

Fig. 24 — Valve Seat Cutter

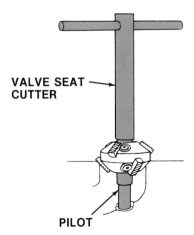

Fig. 26 — Removing Valve Seat

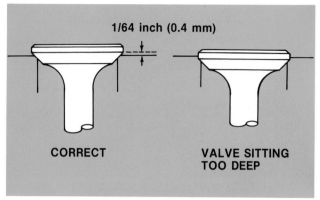

Fig. 27 — Valve and Seat Relationship

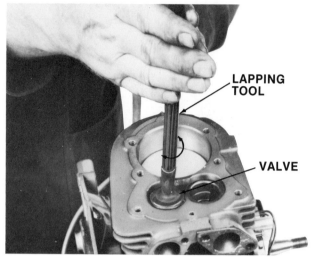

Fig. 29 — Lapping Valves

When the seat has been ground smooth, free of imperfections, and to the recommended width, remove the cutting tool from the bore and insert the valve. The lower edge of the margin should sit approximately 1/64 inch (0.4 mm) above the valve seat face (Fig. 27). If the valve sits too deep or the seat is too wide, replace the mandrel and reduce the top edge of the seat face with a narrowing cutter until the dimension is met. Measure the valve seat with a caliper (Fig. 28).

Another method to check the width and pattern of the valve-to-seat face contact area is to apply bluing to the seat face. Then insert the valve, and turn it very slightly in the seat.

The blue transferred to the valve face should be even and of the recommended contact width. The width will be narrower if an interference fit is used. If the blue transferred is not even, the valve stem may be bent or the seat distorted and replacement or further work is necessary.

VALVE LAPPING

Lapping the valve to the seat after grinding may be necessary, depending on a number of factors.

The first is the accuracy of the stone used for grinding. If a diamond tool was used to keep the stone face clean and true, the seat should be true.

Second, if the stone in the valve refacer was trued, the valve surface should match the seat without further work. Sometimes these conditions are not met in the field, and very often in engine repair, carbide cutters are used rather than high speed revolving stones. Such cutters tend to leave small chatter marks.

In this case, it may be better to coat the valve face lightly with fine lapping compound and use a standard, oscillating lapping tool to rotate the valve to produce a light gray finish on both seating surfaces (Fig. 29). Never overlap since it will only produce a rounded, grooved seat.

Fig. 28 — Measuring Valve Seat

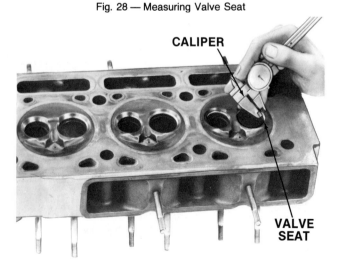

Fig. 30 — Checking Valve Spring with Square

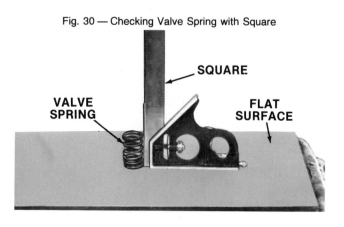

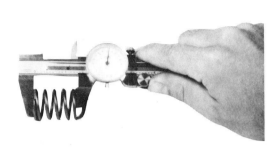

Fig. 31 — Checking Valve Spring Length

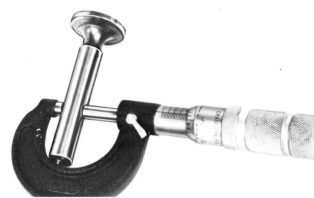

Fig. 33 — Measuring Cam Follower Stem

Valve Springs

Valve springs on small engines are usually tested only for distortion and proper length.

Check distortion by placing the valve spring on end on a flat surface against a square (Fig. 30). The spring should meet the square for its full length throughout its entire circumference. If it does not, it is distorted and should be replaced since its action will cause uneven wear of the valve and guide.

Spring length is checked with a caliper or graduated square and flat surface (Fig. 31). The length should be within the tolerances given in the manual.

Although often overlooked in small engine repair, check the valve spring tension before replacing a used spring especially for higher output and more heavily loaded engines. A simple lever and scale tester will show if the spring tension is within the usual 10 percent limits (Fig. 32). It will also often reveal a spring about to break because of an unseen crack.

Note also whether the coils are evenly spaced. Many valve springs are wound with several turns more closely spaced than the others and these are called "damper" coils. Install them so they are up or down as in the original installation.

Also examine the valve spring cups for excessive wear. If the metal is badly worn, the cups should be replaced.

CAM FOLLOWERS, PUSH RODS, AND ROCKER ARMS

If the valve train contains cam followers, push rods, and rocker arms, check their dimensions. Measure the stem of the cam follower with a micrometer (Fig. 33) and the inside diameter of the cam follower bore in the block with a telescoping gauge.

To check for bent push rods, lay them on a flat surface and use a feeler gauge (Fig. 34). Also, check push rod length. If either measurement does not fall within the limits specified, replace the push rod.

Fig. 32 — Measuring Valve Spring Tension

Fig. 34 — Check for Bent Push Rod

Fig. 35 — Measure Rocker Arm Shaft

Fig. 37 — Remove Camshaft Bearing Retainer Screw

Measure the rocker arm shaft (Fig. 35) and rocker arm bushing bore (Fig. 36). Replace any part that is not within the specification.

VALVE ASSEMBLY

Assemble valves in the reverse order of disassembly using the same spring compression tool. Before inserting the valve stem in the guide lubricate the stem with light engine oil. If rubber or neoprene valve seals are used, replace them; never try to reuse them. Assure that the keepers are properly in place and seated before releasing the spring compressor.

CAMSHAFT

Before removing the camshaft, examine the drive gears for timing marks to avoid confusion later. Camshaft removal procedures differ depending on the type of engine being serviced.

For **multiple-cylinder diesel engines**, first remove the timing gear cover, rocker arm cover, and cylinder head. Also remove the camshaft bearing retainer screw (Fig. 37), if one is present.

Then use a magnetic holding tool set to hold cam followers away from the camshaft lobes during removal (Fig. 38). Carefully remove the camshaft from the cylinder block, taking care that the cam lobes do not drag in the bores.

For **single-cylinder gasoline engines,** remove the end wall section of the engine crankcase or cylinder block. Remove the oil slinger and governor assembly (Fig. 39). Rotate the crankshaft until the timing marks on the crankshaft gear and camshaft gear align (Fig. 39) and then remove the camshaft.

After removal examine the cam lobes for wear and damage and replace the shaft if necessary. Measure height of lobes and bearing surfaces of the camshaft with a micrometer (Fig. 40).

Fig. 38 — Magnetic Holding Tool Set

Fig. 36 — Measuring Rocker Arm Bushing

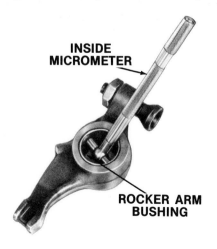

INSIDE MICROMETER

ROCKER ARM BUSHING

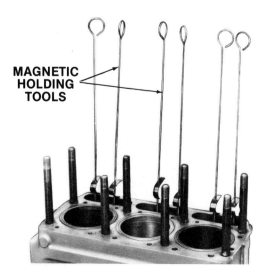

MAGNETIC HOLDING TOOLS

Fig. 39 — Remove Oil Slinger and Governor Assembly

Installation

Clean the cam and bearing surface and reinstall the camshaft.

Camshaft timing is obtained by aligning timing marks on the timing gears with a mark on a gear with which it meshes (See Fig. 39). Examine and adjust camshaft end play if end play is adjustable. See the technical manual for the correct procedure to use. Different engines may use thrust washers, spring loads, and bushings to control end play.

VALVE CLEARANCE

Valve clearance is the gap between the valve stem tip and rocker arm or cam follower that exists when the piston at top dead center (TDC) of its compression stroke on each cylinder. In this position, the intake and exhaust valves are closed. The valves must be closed for the gap to exist in the valve train.

Most engines have a timing mark on the flywheel to indicate TDC. All multiple-cylinder engines contain this mark to find TDC of the No. 1 cylinder. The technical manual gives the engine firing order and tells how much to rotate the flywheel to find TDC of the next cylinder.

For engines without this mark, methods for determining TDC will be covered in the following discussion.

First, however, it is necessary to determine the compression stroke of each cylinder. This is done after the engine is assembled.

Find the compression stroke of a cylinder by first removing the spark plug or injection nozzle for that cylinder. Hold your thumb over spark plug or injector hole while turning the engine by hand until compression is detected. At this point the piston is near TDC.

Now let's discuss the methods for accurately finding TDC on engines of different designs.

Measure the crankcase or cylinder block bearing surfaces with an inside micrometer or telescoping gauge. Check size and for out-of-round conditions. Replace and rebore the crankcase cam bushings if needed assuring that the oil holes in the bushing and block are aligned to allow for proper lubrication. Some precision bushings won't require boring.

When reaming or reboring a bushing, it is important to maintain the original gear centers between a camshaft gear and other gears with which it meshes. If the centers are too close, the gears will run too tight. If the centers are too far apart the gears will run loose.

Before attempting to ream or rebore, measure the existing gear backlash. Remember, if the old assembly seems to have loose gears, it may be because the camshaft bushing, crankshaft bearing, or idler spindle is loose. You must determine these factors before making a judgement on your next step. Some manufacturers provide oversized and undersized gears to compensate for center distances that may have been altered by reaming or reboring a bushing.

Fig. 40 — Measuring Cam Lobes

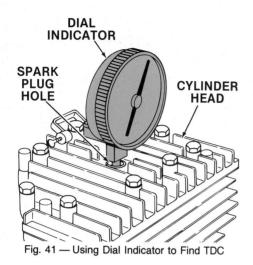

Fig. 41 — Using Dial Indicator to Find TDC

I-Head Engines (without TDC mark)

When the spark plug or injection nozzle hole is directly above the piston, a dial indicator may be used to find TDC. Insert the dial indicator in the hole to determine maximum travel of the piston (Fig. 41).

L-Head Engines (without TDC mark)

When the spark plug or injection nozzle hole is not positioned directly above the piston, find TDC by observing the camshaft and cam follower while rotating the flywheel back and forth.

When the flywheel is rotated slightly back and forth, the camshaft turns and alternately lifts (Fig. 42, left), lowers (Fig. 42, right), and again lifts (Fig. 42, center) the cam follower and valve.

Mark the position of the flywheel, in both directions of rotation, when the cam follower begins to move up (Fig. 42). The point halfway between these two marks is where the piston is at TDC and where the gap in the valve train is the greatest. Measure valve clearance when the flywheel is in this position.

Engines With Rocker Arm And Push Rod

Engines with rocker arms and push rods almost always have more than one cylinder. Therefore, as discussed earlier, there is a mark on the flywheel to use to locate TDC of the No. 1 cylinder on its compression stroke.

The gap where valve clearance is measured is between the valve stem tip and the rocker arm (Fig. 43).

ADJUSTING VALVE CLEARANCE

The clearance is set in many different ways. The method used depends on the design of the engine and valve train. For all designs, make sure the flywheel does not rotate while adjusting valve clearance for

Fig. 42 — Rotating Flywheel to Find TDC

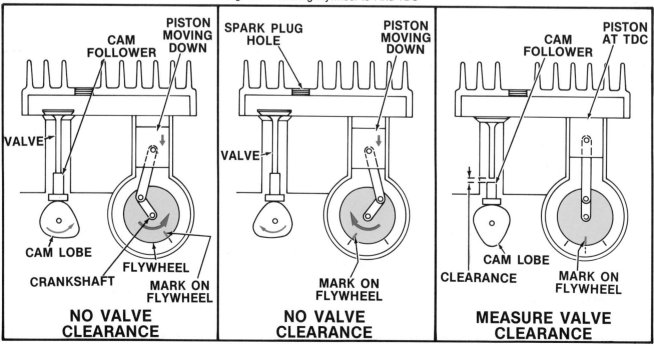

NO VALVE CLEARANCE   NO VALVE CLEARANCE   MEASURE VALVE CLEARANCE

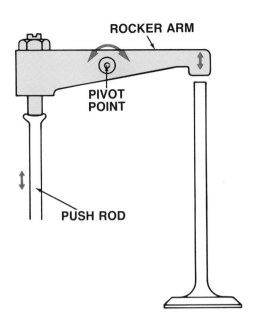

Fig. 43 — Clearance Between Valve Stem Tip and Rocker Arm

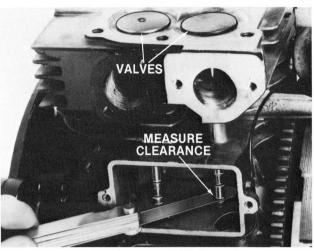

Fig. 45 — Measuring Valve Clearance

each cylinder. Two methods which are commonly used require the technician to:

• *Turn an adjusting screw on the rocker arm or cam follower (Fig. 44, left)*

• *Grind the stem tip or cup (Fig. 44, center)*

• *Use shims between the valve stem tip and cam followers cup (Fig. 44, right)*

Consult the technical manual for the method to use to obtain the proper valve clearance as well as the

amount of clearance required. Use a feeler gauge to measure the clearance (Fig. 45). In Fig. 45, the clearance is measured between the cam follower and the valve stem tip.

On multiple-cylinder engines, valve clearance must be set for each cylinder individually. Make sure that the piston is at TDC on the compression stroke for each cylinder.

CYLINDER BLOCK AND CYLINDERS

CLEANING

Recondition the cylinder block by first stripping the block of all engine components and carefully scraping away all remnants of old gasket material. Further cleaning depends on the block material and whether the engine is air or liquid cooled.

Fig. 44 — Methods of Adjusting Valve Clearance

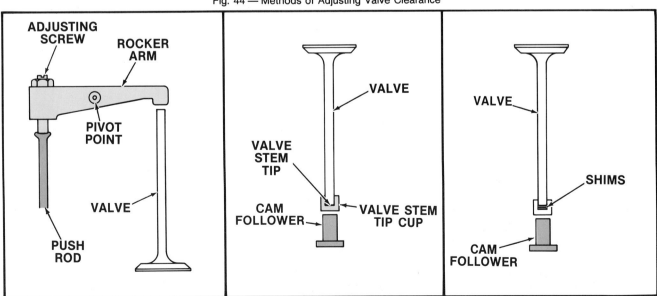

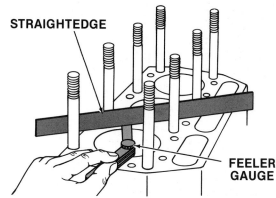

STRAIGHTEDGE

FEELER GAUGE

Fig. 46 — Checking Cylinder Block for Flat Mating Surface

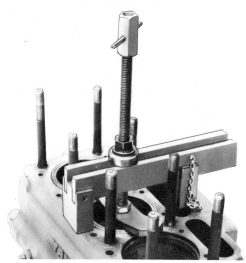

Fig. 47 — Removing Cylinder Liner

Normally clean cast-iron blocks by soaking and agitating the entire block in a vat of hot commercial heavy duty cleaner. Then thoroughly rinse the block with very hot water or steam.

Rinse all traces of the cleaner from the engine. Use suitable brushes to clean oil and coolant passages. Exceptionally heavy scale deposits in coolant passages may require a bath in a commercial pickling acid prior to the alkaline cleaning. Since many of these solutions will damage aluminum, replace aluminum pieces, such as bushings, after such a cleaning.

Normally clean aluminum blocks in a "cold tank" containing solvents which do not damage aluminum. A sequence of soaking, brushing to remove residue, and resoaking and brushing is recommended before rinsing with hot water. Wash all traces of the solvent from the block. Cold tank solvents also work quite well on cast-iron blocks, unless descaling of coolant passages is necessary. These solvents have the advantage of not affecting aluminum bearing surfaces.

Make sure the solvents, acids, or alkalines that are used to clean the engine will not damage the bearing surfaces. If those surfaces are damaged, new bearings must be line bored on the exact same centers as the old ones. Many service departments are not equipped to provide this service.

INSPECTION

Inspect the cylinder block for cracks and discard the block if any are found. Check threaded holes for burrs and thread damage. Restore damaged threads by tapping. Replace stripped threads with commercially available inserts. Machined surfaces must be free of scratches and burrs. Use a file hone to remove scratches and burrs from machined surfaces. Deep scratches in critical interface areas may require rema-

chining the surface. Scratches may cause early gasket or component failure.

Check the cylinder head mating surface for flatness in the same manner as used on cylinder heads (Fig. 46). A warped surface should be remachined or discarded if severely warped.

Cylinder liners may be cast-in-block or a removable wet-sleeve-in-block as discussed in Chapter 2.

Inspect cylinder bores for deep scratches, scores, or gouges. Damage from broken piston rings or other metal fragments will require honing or boring to oversize by one or more piston sizes. These procedures will be discussed later in this chapter. A new short block is often an economical alternative to reboring light engines with cast-in-block cylinders.

If a removable cylinder liner requires servicing or replacement, first remove the old cylinder liner (Fig. 47) for inspection.

Reboring cylinders, requires special tools and equipment. Cylinders requiring this type of repair are normally sent to a specialty shop or back to the factory.

If the cylinder does not show extensive gouging, determine the amount and type of wear. Use an inside micrometer (Fig. 48) or telescoping bore gauge to measure the diameter of the cylinder at the top and bottom of the piston ring travel.

Measure the diameter twice at each location—once parallel to the axis of the crankshaft and again at right

Fig. 48 — Measuring Cylinder Bore

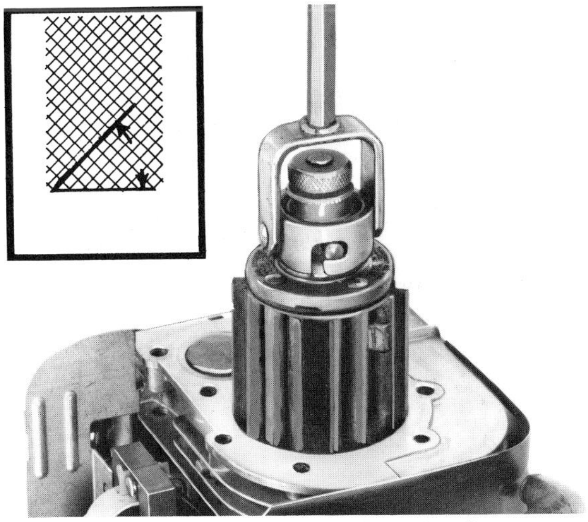

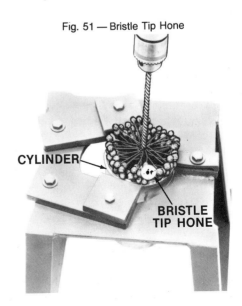

Fig. 50 — Hone with Deglazing Stone

angles to it. These two measurements show the amount the cylinder is out-of-round at that location. Check at both top and bottom to establish the amount of cylinder taper.

Taper and out-of-round measurements must not exceed the manufacturer's specification. Bore diameter measurements over the maximum require that the cylinder be rebored or replaced.

Depending on the amount of reboring required, a special set of oversized piston rings or a new piston may be required. Oversized piston rings are most commonly available in sizes 0.010, 0.020, 0.030, and 0.040 inch (0.25, 0.5, 0.76, and 1.02 mm) oversized. However, some engines may have rings as small as 0.005 inch (0.13 mm) or as large as 0.060 (1.5 mm) oversized. Refer to the technical manual. If wear is within the tolerance limits the cylinder can be reconditioned by:

- **Ridge reaming**

- **Honing or deglazing**

Fig. 49 — Removing Cylinder Ridge

Ridge Reaming

Reciprocation piston movement produces a cylinder wear pattern concentrated at the top of the ring travel. An area of worn metal below and a ridge of unworn metal above forms. If this ridge is not removed, the piston ring could catch on the ridge and break the ring or damage the piston.

Remove this ridge from the cylinder with a tool called a ridge reamer (Fig. 49). Place the reamer with its running surface flat on the top of the cylinder and its cutters extending into the ridge area. Adjust the cutters to take a light cut of the ridge when the reamer is rotated. Use successive light cuts and do not cut into the piston ring travel area. After the ridge is removed the cylinder is ready for honing.

Ridge reaming is usually done only on cast-in-block cylinders. Wet-sleeve-in-block cylinders with a ridge should be replaced.

Fig. 51 — Bristle Tip Hone

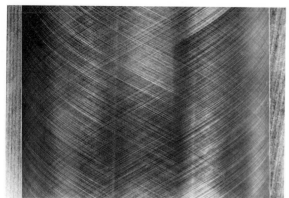

Fig. 52 — Cross-hatch Finish in Cylinder Bore

Honing or Deglazing

A cylinder hone or deglazing tool is a tool for removing high spots from the cylinder and for establishing a surface finish conducive to proper ring seating. Two types of hones are available. They may contain a stone for deglazing (Fig. 50) or brushes with coated bristle tips (Fig. 51). The grit size of the hone is selected to produce the recommended surface on the cylinder wall. Hones produced by various manufacturers will have different lubricant requirements. Follow the instructions which come with the hone.

Fit the hone to the narrowest part of the bore so it drags freely up and down the bore when it isn't rotating. After the hone is fitted, start it rotating and move it up and down the full length of the bore. Maintain a constant up and down movement. High spots will show up as areas of increased drag.

Keep the hone well lubricated, if it is a wet hone. Wire brush the cutting surfaces occasionally to prevent them from filling up with metal particles.

When the high spots have been reduced, make a final pass up and down the bore using the same constant speed strokes to impart an even, cross-hatched finish to the entire bore (Fig. 52). This finish will hold oil in the bore surface during piston ring beak in. Avoid overhoning. Remove only as much metal as necessary to provide a smooth, cross-hatched bore.

After honing, recheck the bore diameter as previously described. Check the readings against the manufacturer's tolerances and follow technical manual instructions regarding the use of oversized piston rings.

Wipe reconditioned cylinder walls free of all abrasives, wash with hot water and detergent, and wipe clean. Wash and wipe until a clean white rag can be wiped firmly in the bore and show no discoloration. When the cylinder is clean, oil it with clean, light engine oil.

INSTALLATION

When installing a new or used cylinder liner, clean all contact surfaces on both the block and liner. Always replace the seals, gaskets, and "O" rings if the cylinder has been moved or removed. Shifting the position of a wet cylinder liner may destroy the seals. Turning the engine over, for inspection purposes, with the head off and the cylinder unrestrained can cause this damage. Restrain the sleeves to prevent shifting during service operation.

Fig. 53 — Remove Engine Covers to Access Connecting Rod Caps

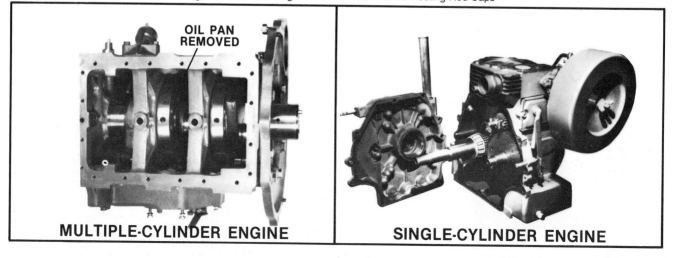

OIL PAN REMOVED

MULTIPLE-CYLINDER ENGINE

SINGLE-CYLINDER ENGINE

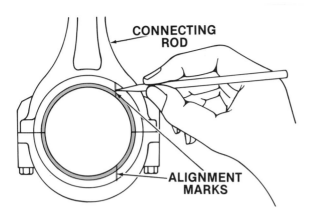

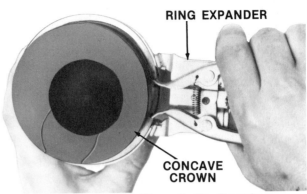

Fig. 55 — Removing Rings with Ring Expander

Fig. 54 — Marking Connecting Rod, Cap, and Bearings

PISTON AND CONNECTING ROD

DISASSEMBLY

To remove the connecting rod and piston, first remove the oil pan (Fig. 53, left) to expose the connecting rod cap. Some engines, especially single-cylinder gasoline engines, do not have a removable oil pan. Instead they contain a removable end wall of the crankcase or cylinder block (Fig. 53, right) for gaining access to the connecting rod cap. The cylinder head for both styles of engines must also be removed before the piston and connecting rod can be removed.

Remove the capscrews and the connecting rod cap along with the crankpin (connecting rod) bearings that fit around the crankpin. Some slash-lubricated engines contain an oil dipper which is part of the connecting rod cap. The dipper is removed with the cap. Remove the piston and connecting rod by pushing them out through the cylinder and the top of the cylinder block.

After removing each connecting rod and cap, check to see if they contain aligning marks. If not, use a punch to place a light alignment mark on the connecting rod and cap as well as on each bearing half. The marks should be adjacent to each other but on opposite sides of the split point (Fig. 54). Realign the marks during assembly.

Also mark the connecting rod, cap, and bearings so that the parts are not mixed. All parts must be assembled around the same crankpin from which they were disassembled.

A piston assembly must be disassembled before it can be effectively cleaned. Remove the rings, and discard them. Never reuse piston rings in a 4-cycle engine. Do not break the old rings to remove them. This may damage the ring grooves. Remove the rings with a ring expander (Fig. 55).

NOTE: Some diesel engine pistons (as in Fig. 55) contain a concave shaped crown. This is to make the diesel fuel swirl when it is injected so it mixes more completely with the air and gives better combustion.

Fig. 56 — Removing Piston Pin and Bushing

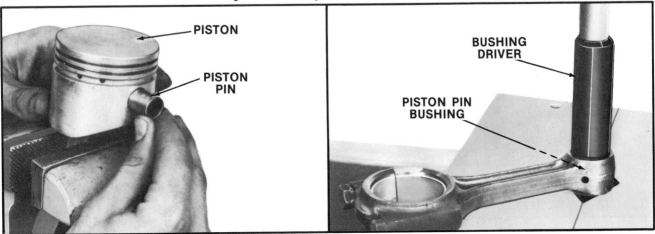

Remove the piston pin from the connecting rod by removing the pin retainers and pushing the pin out with a finger (Fig. 56, left) or with a piston pin tool. Keep the pin with its original piston to avoid unnecessary refitting later. Also remove the piston pin bushing from the eye of the connecting rod (Fig. 56, right).

CLEANING

Clean pistons, connecting rods and bearings by soaking in a solvent designed for the purpose. Follow the directions which come with the solvent. After the solvent bath, brush the piston with a stiff bristle brush (never a wire brush or wheel) and then wash with water. If all the carbon has not been removed (pay special attention to the ring grooves) resoak in solvent and repeat the cleaning.

In some cases a light scraping may be used to remove carbon; and never enough to remove metal from the piston. Glass bead or vapor blasting may also be used to clean pistons, but **both methods require extreme care**. The blast will quickly distort the corners of grooves and lands causing a poor ring seal when the engine is reassembled. A ring groove cleaner (Fig. 57) may be used for additional cleaning of the grooves.

PISTON INSPECTION

Inspect the cleaned piston for cracks, burned or overheated areas, and scoring or scuffing. Replace cracked, heat damaged, or severely scored pistons.

Check the ring grooves for wear by inserting the edge of a new ring into the groove and checking the remaining gap with a feeler gauge (Fig. 58). Check grooves of tapered ring with a special gauge available

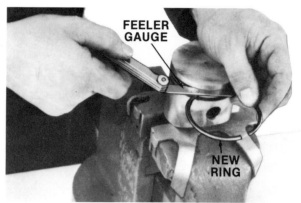

Fig. 58 — Checking for Ring Groove Wear

from the manufacturer. Special tools are required to regroove pistons. A spacer is then inserted above the ring. This practice is not common on small engines. Check the manual to find if the manufacturer recommends regrooving or replacing pistons that do not meet the specification.

Check piston diameter for excessive wear by measuring at the top and bottom of the skirt at right angles to the pin bore (Fig. 59). These measurements may indicate a slightly elliptical shape. This characteristic is normal since different areas of the piston expand at different rates as the piston warms up. Then check the measurement against those in the manual. Reject the piston if the wear is excessive.

The piston may be within limits, but still be too small for the bore if the bore has been reconditioned. Check the diameter of the cylinder at right angles to the crankshaft at the lower end of the bore. This diameter, compared to the readings taken from the lower end of the piston skirt will give the total piston clearance.

Fig. 57 — Ring Groove Cleaner

RING GROOVE CLEANER

Fig. 59 — Measuring Piston

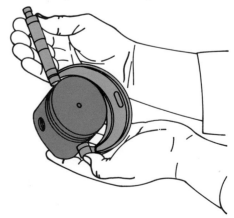

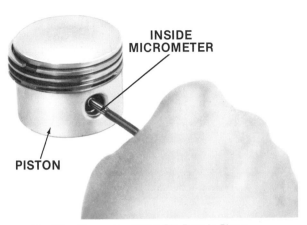

Fig. 60 — Measuring Piston Pin Bore in Piston

Fig. 62 — Oil Consumption — How Piston Rings Pump Oil into Cylinders where it is Burned

Check these findings against the specifications to determine if piston replacement or oversized piston rings are required. If suitable bore gauges and micrometers are not available, insert thin feeler stock between the piston and cylinder wall at the bottom of the cylinder and use a light scale to measure the force needed to withdraw it. Many manufacturers specify piston clearance in this way.

Examine the piston pin bore for scratching or excessive wear. Measure the bore in two dimensions at right angles to each other for size and out-of-round condition (Fig. 60). Measure the piston pin and inspect in the same way (Fig. 61). Compare the measurements and consult the specifications for out-of-tolerance conditions. Check the connecting rod bearing the same way. If the clearances are excessive, replace the bearing and ream or hone the new bearing to the proper size.

ANALYZING PISTON FAILURES

The early signs of impending piston failure are excessive blowby and oil consumption.

Blowby is the loss of combustion gases past the piston and rings and into the crankcase. No piston forms a perfect seal, therefore, small amounts of blowby are a normal part of engine operation. Crankcase breathers ventilate these combustion gases. Excessive blowby, however, causes engine power loss, and early contamination and failure of the lubricating oil.

Slight oil consumption is also normal and small amounts of oil consumed in cylinder lubrication prevent wear. Excessive oil consumption, however, can cause spark plug fouling, incomplete combustion of fuel, and smokey exhaust. One common source of oil consumption is the pumping action caused by worn rings and grooves as shown in Fig. 62.

Some piston and ring failures are the result of normal wear. Others reflect other factors, such as oil pump failure, within the engine. If only the piston is repaired, the problem may recur. Examine the pistons at the time of disassembly to determine the cause of failure.

Fig. 61 — Measuring Piston Pin

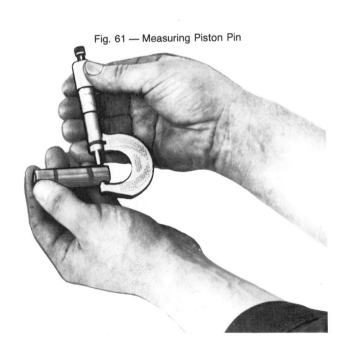

Fig. 63 — Piston Rings Stuck and Broken because of Oil, Fuel, or Carbon Build-Up

Fig. 64 — Piston and Rings Worn by Abrasives

Common piston failure are:

• **Stuck rings or plugged oil ring**

• **Top ring broken or top groove worn**

• **Abrasive wear**

• **Scuffing and scouring**

• **Corrosion**

• **Physical damage**

The failures just mentioned are usually the result of some other malfunction in the engine. For example, plugged oil rings may have been caused by dirt entering the lubrication system. If this is the case, examine the engine to determine where dirt may have entered. If only the piston and rings are replaced, without repairing the source of oil contamination, a repeat failure is likely.

An in-depth failure analysis is beyond the scope of this manual. We will, however, give possible causes for the above failures to assist you in knowing where to begin to look for causes.

Stuck rings and plugged oil rings (Fig. 63) may result from hardened deposits of oil, fuel, or carbon building up in the area between ring and groove. Improper or contaminated lubricating oil is a major cause. Other causes are top groove wear, cylinder wear or distortion, cooling system failure, or lubrication system failure. ›

The top ring and groove must seal to each other and to the cylinder wall. A new ring installed in a worn ring groove cannot seal properly. Other causes of ring and

groove failure are abrasives in the combustion chamber, combustion knock, a residual cylinder ridge after rebuild, inadequate ring gaps, or the use of the wrong sized ring.

Abrasive wear (Fig. 64) is usually due to sand, dirt, or other abrasives that enter the combustion chamber. This may be due to a faulty or missing air cleaner or through a manifold leak. Abrasives produce dull vertical scratches on the piston rings which may extend to the piston. Abrasive material may also enter the engine during overhaul or when adding or changing oil.

Scuffing and scoring (Fig. 65) usually result when two metal surfaces rub together with insufficient lubrication. Cooling system failure or hot spots on the cylinder wall due to cylinder distortion may also cause these conditions.

Fig. 65 — Piston and Rings Scuffed and Scored

Fig. 66 — Diagonal Wear on Piston from Bent Connecting Rod

Physical damage to a piston may be caused by:

- **Misalignment or excessive looseness**
- **Debris (including loose parts) in the cylinder**
- **Detonation or preignition**
- **Improper installation after resizing cylinder**

Misalignment of engine components may result from excessive crankshaft, end play, a bent connecting rod (Fig. 66), misaligned cylinder, or cocked piston pin.

Debris between cylinder and piston or piston head and cylinder head can cause piston, cylinder, and head damage. Common examples are loose screws and other mechanical parts which have made their way through the intake valve, broken valve heads or seat inserts and broken or loose piston pin retainers and pins (Fig. 67). Another problem causing the same type

Fig. 67 — Piston Damaged from Piston Pin Retainer Coming Loose

Fig. 68 — Piston Damaged by Detonation

of damage is a mistimed or misadjusted valve. Severe overspeeding may cause valve and piston interference in overhead valve engines.

Detonation damage (Fig. 68) is caused by explosive combustion of fuel possibly caused by low octane, advanced ignition timing, lean fuel mixture, excessive heat, or too high compression. Detonation is generally a problem only in gasoline engines.

Preignition damage (Fig. 69) can be caused by glowing combustion chamber deposits, a hot exhaust valve, or spark plugs of the wrong heat range. Fuel is ignited while the engine is still on the compression stroke and, if ignited at the wrong time, can break pistons, piston rings, cause blowby and piston burning, or bend connecting rods. Preignition is a condition also normally found only in gasoline engines.

CONNECTING ROD AND CRANKPIN BEARING INSPECTION

Visually inspect the connecting rod for cracks, bends, or twists. An irregular wear pattern on piston pin bushings or crankpin bearing surfaces will generally indicate if the connecting rod is bent or twisted.

Fig. 69 — Piston Damaged by Preignition

Fig. 70 — Bearing Damaged by Oil Starvation

Fig. 71 — Corrosion from Acid Formation in Oil

Examine the bearing inserts for signs of wear. The illustrations show examples of bearings which have been damaged by:

- **Oil starvation (Fig. 70)**
- **Corrosion (Fig. 71)**
- **Dirt (Fig. 72)**
- **A bent connecting rod (Fig. 73)**
- **Tapered crank journals (Fig. 74)**
- **Metal fatigue due to overheating (Fig. 75)**

In each case, examine the crankpin or journal at the same time for signs of obvious wear, metal transfer, or heat buildup. Also note oil groove ridging around the circumference of journals and crankpins.

Fig. 72 — Dirt Embedded in Bearing

Connecting rods without bearing inserts are common on light engines. The rod end and cap act as the bearing surface. The symptoms of bearing failure shown still hold true.

Fig. 73 — Excessive Wear Caused by a Bent Connecting Rod

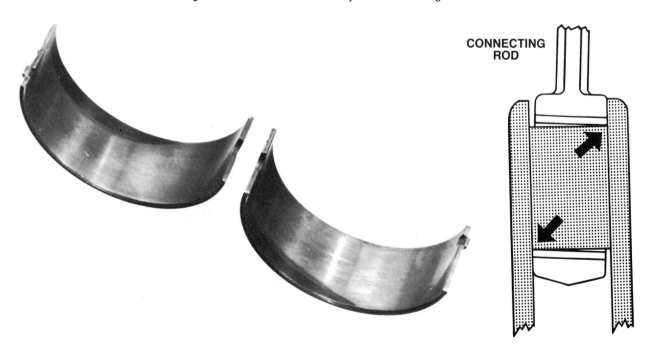

CONNECTING ROD

Fig. 74 — Wear on One Edge of Bearing Caused by Tapered Journals

Fig. 75 — Bearing Fatigue Caused by Overloading and Heat

Check the piston pin bushing for size and taper by measuring both pin and the bore of the piston pin bushing (Fig. 76). Compare the measurements with the manufacturer's specifications. Normally, these clearances are not a problem on small engines and the following quick check of the way the pin fits the bushing can give a good indication of the overall condition.

If the pin:

• *Enters the bushing easily at one end but fits tight at the other the bore is tapered.*

• *Enters easily at both ends but is tight at the center the bore is bell-mouthed.*

• *Enters and passes through the bore with equal pressure the bore is satisfactory.*

Check bushing and pin for galling or scratches and for signs of corrosion.

Fig. 76 — Measuring Bore of Piston Pin Bushing

Fig. 77 — Measuring Crankpin Bearing Bore and Crankpin

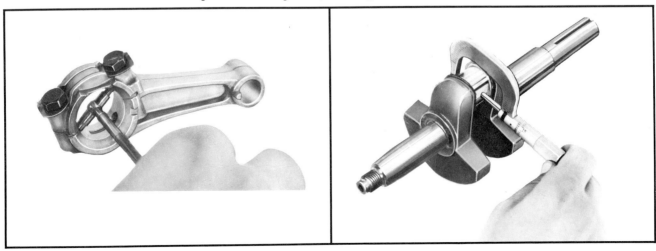

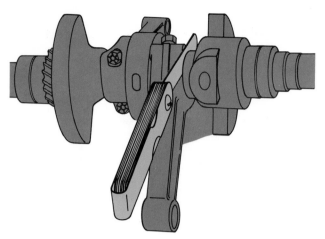

Fig. 78 — Measuring Side Clearance of Connecting Rod

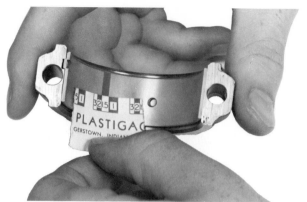

Fig. 80 — Determining Bearing Clearance with Plastigage

If the bushing falls within measured tolerances, is parallel, and has a smooth bearing surface it is acceptable for reuse.

Check the crankpin bearing surface in the connecting rod head for size and roundness. To do this:

1. Install a new set of inserts if inserts are used

2. Assemble the head using the correct torque values on the capscrews. Then use an inside micrometer to measure the dimensions at several points on the inside of the bearing surface (Fig. 77, left). Compare these measurements with those of the crankpin outside diameter (Fig. 77, right) and consult the technical manual for proper clearances.

For connecting rods which have no inserts, check for taper in the bore in the same manner as described for checking crankpins. Reassemble the connecting rod to the crankpin.

Use a feeler gauge to check side clearance of the rod between the cheeks of the crankshaft throws (Fig. 78).

Fig. 79 — Placing Pastigage on Bearing Insert

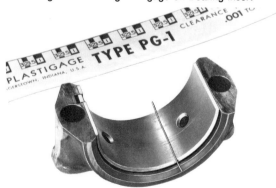

Another method of checking bearing fit is by the use of Plastigage. Plastigage will indicate the overall clearance but will not show whether wear is in the bearing or in the crankpin. To use Plastigage:

1. Place a piece of Plastigage the length of the bearing surface about 1/4 inch (6 mm) off center (Fig. 79).

2. Install the connecting rod on the crankpin and tighten the capscrews to the recommended torque.

3. Remove the bearing cap. The flattened piece of Plastigage will adhere to either the bearing surface or the crankpin.

4. Compare the width of the flattened Plastigage at its widest point with graduations on the Plastigage envelope (Fig. 80). The graduation which matches the width of the material indicates the total clearance in thousandths of an inch (mm).

Any taper in the clearance will also show up as a taper in the width of the crushed Plastigage and can be determined as the difference in clearance between the two ends.

REASSEMBLY OF PISTON AND PISTON RINGS

New piston rings come in sets and generally have instructions enclosed for determining the top of the ring and method of installation. Some general practices relating to ring installation include:

• *Make sure the rings are the right type and size. As a precaution measure the ring end gap.*

• *Always install rings top side up.*

• *Always use a ring expander. Never try to flex rings on by hand.*

• *Don't over expand rings. They'll permanently distort.*

• *Make absolutely sure the groove is clean and free of carbon deposits.*

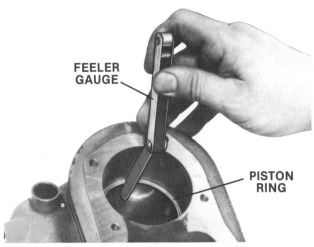

Fig. 81 — Checking Ring End Gap

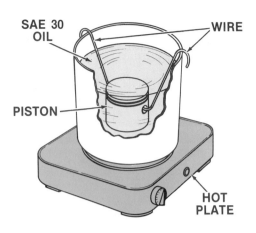

Fig. 82 — Heating Piston before Piston Pin Installation

Most of these precautions are self explanatory, but measuring end gap is an important and often misapplied procedure.

To measure end gap, take the top compression ring from the new set and compress it slightly to fit it into the cylinder bore. Use the piston to push the ring into the bottom area of ring travel and assure that the ring is square with the bore.

Using feeler gauges, determine the gap, or distance between the two ends of the ring (Fig. 81). Compare this dimension to the manufacturer's recommendation for the engine. If the gap is too great, and the bore size within limits, oversized rings are probably needed.

NOTE: Always make sure the ring is in the area of piston travel. A ring placed in the area above the travel may show an inaccurate reading due to cylinder top "belling" caused by ridge reaming. This belling is undesirable, but if it does not extend into the area of piston travel it will not greatly affect the performance or life of the engine.

After the rings are installed on the piston, the piston pin and connecting rod can be installed. (The order of these procedures is optional). The connecting rod should, of course, be cleaned, inspected, and its bushing sized to the piston pin prior to this assembly.

Assemble the piston pin in the connecting rod in the same manner used for disassembly. Make sure any lubrication holes are properly aligned. It may be required that the piston be heated before the piston pin can be installed. If heat is required, heat the piston in a container filled with oil. Suspend the piston in the container by bending the ends of a wire and hooking one end in the piston pin bore and the other end over the edge of the container (Fig. 82).

The piston should not touch the sides or bottom of the container. These surfaces get hotter when the oil is heated (as described below) and the piston could be warped if it were to touch the side.

Place the container on a hot plate and begin to heat the oil. When the oil begins to smoke, remove the piston and push in the piston pin. Make sure the pin is properly positioned before the piston cools.

Apply light oil to the bore before assembling the piston pin to the connecting rod. Make sure that all keepers and locks are in place and properly installed.

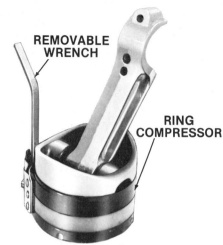

Fig. 83 — Installing Ring Compressor

Fig. 84 — Installing Piston Assembly in Cylinder Block

INSTALLATION

To install the connecting rod and piston in the cylinder bore:

1. Coat piston and rings with light oil.

2. Position the ring compressor even with the top edge of the cylinder wall.

3. Tighten the ring compressor until the rings are completely compressed (Fig. 83). Overtightening will cause the piston to stick in the compressor.

4. Insert the connecting rod and the piston into the bore until the ring compressor is tight against the top of the bore. Take care that the connecting rod does not damage the cylinder liner bore during installation.

5. Push the piston completely into the bore while holding the ring compressor firmly in position (Fig. 84). **Never strike the top of the piston to force it into the bore.**

6. Gather the crankpin bearings and connecting rod cap that are to be assembled with the connecting rod. Check identification marks to make sure the same parts are used with the same connecting rod from which they were disassembled.

7. Place lubricating oil on the crankpin bearings. Assemble them into the bore of the connecting rod and cap and assemble the components around the crankpin (Fig. 85). Check the marks to make sure the parts are aligned properly.

8. Install capscrews and tighten to specified torque.

CRANKSHAFT, BEARINGS, AND FLYWHEEL

Preliminary visual inspection of parts during disassembly is especially important in the case of crankshafts and connecting rods. Unless the service department is equipped with special tools and machines for inspection purposes (which are normally found only in large engine or machine shops) the preliminary inspection may be the service technician's only opportunity to determine the cause of engine failure.

The vertical crankshaft engines used on walk-behind rotary lawn mowers require a special check prior to disassembly. Because the blade is mounted directly to the crankshaft with no intermediate drives, the crankshaft may bend when the mower hits rocks or debris.

Check for this type of damage while the engine is still on the mower.

Fig. 85 — Assembling Connecting Rod and Cap Around Crankpin

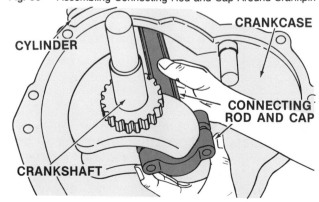

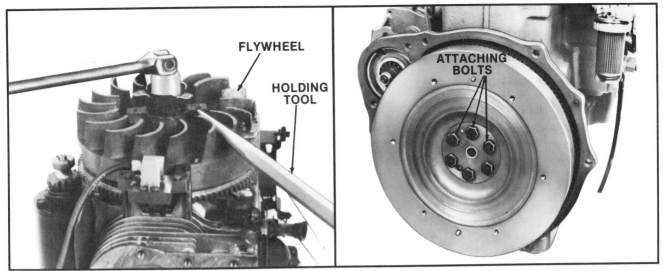

Fig. 86 — Removing Bolts that Hold Flywheel

Caution: To avoid fire, empty the fuel tank before performing the following procedures.

Tip the lawn mower on its side to view the end of the crankshaft to which the mower blade is attached.

Use the pull starter to turn the engine over slowly (with spark plug out and plug wire taped) while watching the end of the crankshaft. If the shaft end turns in a circle instead of rotating on center, the shaft is bent. If the end wobbles through the circle, the bearings are probably worn as well. Obviously, visual inspection will not reveal a bend of only a few thousandths. A dial indicator reading is preferred.

Straightening a bent crankshaft requires special equipment. The straightening process may cause cracks and fatigue spots. Also, it is difficult to get the shaft exactly straight. Replacement is generally less costly and more effective in the long run.

DISASSEMBLY AND INSPECTION

Before removing the crankshaft, remove the flywheel. On larger engines, this only requires the removal of a few bolts and lifting the flywheel off (Fig. 86, left).

On smaller engines the flywheel may be mounted on a tapered or key shaft held in place by a nut threaded onto the end of the crankshaft.

The thread may be right or left hand. Use a special tool to hold the flywheel from turning while freeing the nut (Fig. 86, right). A puller is sometimes required to remove the flywheel after the nut is removed (Fig. 87). Never attempt to loosen a flywheel by placing pry bars

behind it and forcing if off. This procedure can damage the flywheel, crankshaft, the cylinder block, and the ignition components behind the flywheel.

Clean the flywheel and examine it for cracks and broken cooling fins (if so equipped). Replace the flywheel if these conditions exist.

Check the starter ring gear (if so equipped) for broken or damaged teeth. Replace the ring gear on some flywheels by cutting off the old ring gear. A new ring gear can be installed by heating it and dropping it squarely into position on the flywheel. Take care not to overheat. Assure that the gear bevels face the proper direction and that the gear is fully seated before it cools. Most small engines do not have replaceable ring gears and the flywheel must be replaced.

Fig. 87 — Using Puller to Remove Flywheel

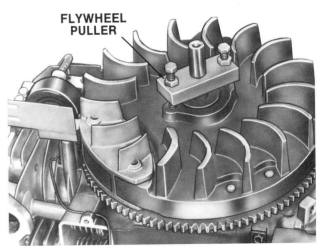

Fig. 88 — Removing Crankshaft Pulley

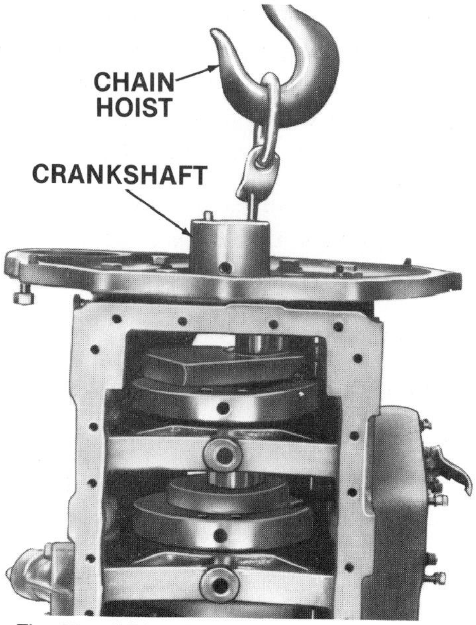

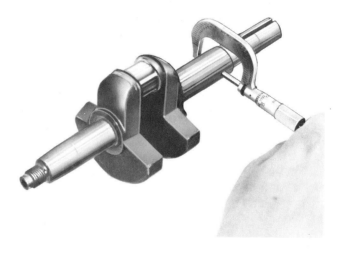

Fig. 90 — Lifting Crankshaft from Cylinder Block

For **multiple cylinder engines,** first remove the crankshaft pulley (Fig. 88). Next, remove the timing gear cover and then use a gear puller to remove the crankshaft gear (Fig. 89).

Invert the cylinder block so the flywheel end is up. Remove the rear main bearing housing capscrews, if present. Attach a chain hoist to the crankshaft (Fig. 90). Slightly lift the crankshaft so the weight is taken off the center main bearing housing.

Remove any bolts from the center main bearing housings and lift the crankshaft from the cylinder block.

For **single-cylinder engines,** remove the end wall of the crankcase or cylinder block as described in the section CAMSHAFT DISASSEMBLY. After removing the camshaft and disassembling connecting rods from caps, remove the crankshaft.

Once these inspection points are noted, remove the crankshaft, wipe clean, and check with a micrometer. Measure all bearing surfaces for dimensions (Fig. 91), out-of-round (measure twice at 90 degrees on the same centerline), and taper (measure at both ends of the crankpin or journal on the same horizontal centerline).

If any of these dimensions exceed the manufacturer's specification, replace or regrind the crankshaft. Additionally, while oil groove ridging up to .001 inch (.025 mm) can be removed by using fine abrasive cloth soaked in fuel oil (finish with wet crocus cloth) larger ridges will require regrinding.

Fig. 91 — Measuring Bearing Surfaces

Fig. 89 — Removing Crankshaft Gear

These points are valid for engines using capped journals. Other engines, however, using ball bearings, roller bearings, bushings, or crankshafts which run directly in the crankcase metal, may have different criteria for crankshaft acceptance. Causes for rejections can include:

- **Damaged flywheel keyways**
- **Damaged threads**
- **Out-of-round or tapered journals**
- **Scratches in journals or crankpins**
- **Signs of corrosion in the same area.**

Many manuals direct the service technician to replace a crankshaft with such damage.

Engines equipped with bushings or those without supplementary wear surfaces require that both the crankshaft and bearing surfaces or bushings be measured. Refer to technical manual specifications to determine if wear is excessive.

Inspect ball and roller bearings and replace as needed.

Ball bearings are usually preassembled. Check them for loose fit, roughness, or a gritty sound when turning the bearing by hand. Replace them if their condition is in doubt.

Inspect roller bearings, especially the tapered variety, visually. Check for scored bearing surfaces, worn or heat-discolored races, and broken retainers.

The technical manual lists the tools and methods to use to remove and replace ball or roller bearings.

Use magnetic particle tests, fluorescent magnetic particle tests, or penetrate tests to check for cracks in crankshafts and connecting rods. Many service departments are not equipped for this type of testing and the only method available is careful visual inspection. Most small engine manufacturers recommend replacement if cracks are found.

REASSEMBLY

Reassemble the crankshaft into the block in reverse order of disassembly. Before final assembly, however, determine the end play of the crankshaft and adjust if necessary.

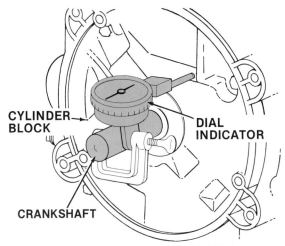

Fig. 92 — Checking Crankshaft End Play

Measure end play at different points on various engines as specified in the technical manual. The points may be between a surface of the crankshaft and some parallel surface of the crankshaft mounting structure. In this case, use a feeler gauge to test for end play. Other engines may require a dial indicator mounted on the block and reading from an external point on the crankshaft (Fig. 92).

The method of end play adjustment can vary also. Most small single-cylinder engines use selective gasket thicknesses between the two parts of the cylinder block to adjust end play. Other engines may use thrust washers or side thrust plates designed into one or more of the main bearing inserts. Follow the recommendations of the technical manual. Be sure that all adjustments are made with new bearings in place.

After determining end play and making necessary adjustments, assemble the crankshaft to the block. Follow the instructions in the technical manual to make sure that all screws and bolts are tightened to the proper torque. If timing to the camshaft, be sure to align the proper marks.

Always replace crankshaft oil seals as a part of major rebuild.

Special sleeves to protect the seal from damage by rough or sharp edges on the crankshaft are available from most manufacturers. Many technical manuals give drawings and dimensions for both sleeves and seal setting tools so the service technician can make up a set. The combination of sleeve and tool have the added advantage of permitting seal installation with a greatly reduced danger of a cocked seal. Many installation tools even set the depth of the seal automatically.

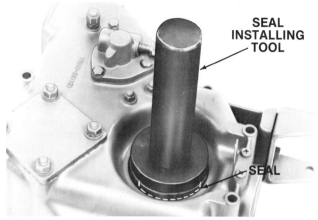

Fig. 93 — Installing Crankshaft Oil Seal

Immediately prior to seal installation, check the seal seat to make sure it is free of dirt, burrs, or other obstructions. Lightly oil the seal and place on the sleeve, place the sleeve over the crankshaft end, and install the setting tool. Drive the seal in place with a seal installing tool (Fig. 93).

Inspect immediately for signs of the seal seated unevenly in its bore or seal damage.

After installing the seals, install the crankshaft gear.

OTHER CONSIDERATIONS DURING CRANK-SHAFT ASSEMBLY

Cam Timing

Many engines have their timing gears or belts located on the outside of the engine block. This usually permits the adjustment of cam timing after the rest of the major components are assembled.

Most smaller engines, however, carry their timing gears inside the cylinder block. This means that the camshaft and crankshaft timing marks must be aligned before the end wall of the cylinder block is assembled (Fig. 94). The technical manual will give detailed instructions as to the requirements of the particular engine.

Before attaching the crankcase or cylinder block end wall, make sure all internal assemblies, such as the following, are in place.

- **Oil pump**
- **Governor**
- **Valve lifter**
- **Oil screens**
- **Fuel injection drives (diesel engine)**

The last check before assembly is to assure that the seal gasket between the end wall and cylinder block is the correct one and properly installed.

Flywheel Installation

Flywheel installation requires that the following procedures be used:

- *Any parts located behind the flywheel must be installed and correctly adjusted (Fig. 95). (The technical manual will cover the details of adjusting ignition points or alternator components. Also see FOS Electrical Systems for Compact Equipment for adjusting procedures.)*

- *Insert the proper key in the crankshaft keyway.*

Fig. 94 — Camshaft and Crankshaft Timing Marks

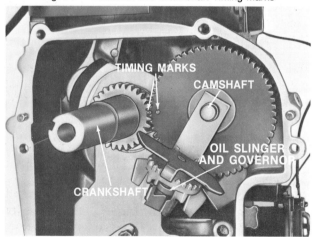

Fig. 95 — Install Parts Behind Flywheel

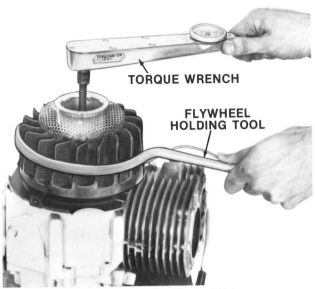

Fig. 96 — Tightening Flywheel Nut

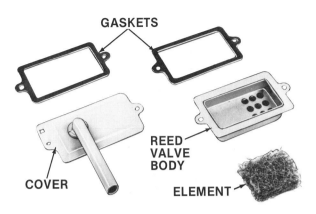

Fig. 97 — Reed Valve Breather Components

• *Align the flywheel keyway with the key and pushing the flywheel onto the crankshaft, taking care that the flywheel is installed squarely on the crankshaft.*

• *Install the flywheel nut and tightening to the specified torque value.*

Any special instructions for flywheel mounting are found in the technical manual along with helpful information on how to hold the crankshaft stable while tightening the flywheel nut (Fig. 96).

Installing the flywheel on larger engines usually involves replacing the flywheel and its securing bolts in reverse order of assembly and tightening the bolts to the specified torque.

Check for flywheel run out by observing a dial indicator reading on the edge of the flywheel. Run out on small tapered-mounted flywheels can indicate the wrong size key in the keyway, a loose retaining nut, or an improperly installed flywheel. On larger flange-mount flywheels run out can indicate burrs on the bolt holes, a distorted flange, or loose or cocked flywheel.

CRANKCASE VENTILATION

Vapors from unburned fuel and water can accumulate in the crankcase and form corrosive contaminants in the oil. The crankcase breather evacuates these vapors under the positive pressure created by the piston on its down stroke. Breathers are of two types:

• *Simple filter (sometimes connected to the carburetor to reduce pollution)*

• *Reed valve breather*

The simple filter breather is an opening in the crankcase fitted with an air filter. It permits vapors to be forced out of the crankcase on the down stroke of the piston and fresh air to be drawn in on the up stroke. If connected to the carburetor the vapors are drawn in with air on the intake stroke and reburned to lower air pollution (this also provides a slight increase in fuel economy).

Cleaning of the filter is the only maintenance required.

A reed valve breather consists of a reed mount, reed, filter and cover (Fig. 97). Positive crankcase pressure in excess of the reeds tension opens the reed and allows vapors to escape. As pressure drops the reed closes again and seals the crankcase. Maintenance on reed valve breathers includes regular cleaning of the filter and occasional replacement of a worn reed (usually at overhaul time).

SUMMARY

When servicing a 4-cycle engine, each component requires careful cleaning, inspection, reconditioning (if needed), and assembly.

After removing the cylinder head, clean the old gasket material from the mating surfaces of the cylinder head and block. Assemble the cylinder head to the cylinder block with a new gasket.

When disassembling the valve train, keep all parts in order so they may be reinstalled in the same position. Inspect valves for wear and failure. If required, the valves, valve guides, and valve seats may be reconditioned. There are several methods to find engine top dead center and adjust valve clearance. The method to use depends on engine design.

Engines may have cast-in-block cylinder liners or wet-sleeve-in-block liners. Either type may be reconditioned by honing. Ridge reaming is usually only practical on cast-in-block cylinders.

Disassemble pistons and connecting rods and mark the parts so they may be reassembled in the same position as removed. Measure the parts for wear and inspect for failed pistons, connecting rods, and crankpin bearings.

Flywheel and crankshaft disassembly depends on whether the engine is single- or multiple-cylinder. After the parts are disassembled, inspect and measure the parts for damage and wear.

Make sure all internal assemblies are in place before assembling the cylinder block. Make sure timing marks on the crankshaft and camshaft are aligned.

CHAPTER 5 REVIEW

1. (True or false) Most engine manufacturers suggest reusing cylinders head gaskets after an engine overhaul.

2. (Multiple choice) One failure that generally does not result in valve seat distortion is:

   A. Valve seats not concentric with valve guide

   B. Cooling system failure

   C. Warped mating surfaces on cylinder head or block

   D. Broken valve spring

3. (Fill in the blank) The process of coating the face with a compound and using a tool to rotate the valve on its seat is called _____ .

4. (Fill in the blanks) Valve clearance is usually measured between the valve stem tip and the _____ _____ or between the valve stem tip and the_____ _____ .

5. Match the term in the left column with its definition in the right column.

   Term
   A. Honing
   B. Reboring
   C. Ridge Reaming

   Definition
   1. Removing high spots from cylinder and leaving a cross-hatched finish in the bore

   2. Reconditioning cylinders to remove gouges, etc. which usually enlarges the entire bore

   3. Removing unworn metal from the cylinder above the ring travel area

6. Why is it necessary to mark connecting rods, caps, and crankpin bearings when they are disassembled from the crankshaft?

7. (True or false) It is normal for the piston skirt to have a slightly elliptical shape.

8. (Multiple choice) The most likely cause of piston scuffing or scoring is:

   A. Detonation or preignition
   B. Cooling system failure
   C. Use of the wrong size ring
   D. Valve clearance out of adjustment

9. (Fill in the blank) _____ is a material placed in the crankpin bearing used to check bearing fit.

10. (True or false) It is not possible to straighten a bent crankshaft.

11. (Fill in the blanks) Two methods of adjusting end play, depending on engine design are by using different thicknesses of _____ or by changing the number of thrust _____ used.

12. (Multiple choice) Select the answer that is **least** likely to be a cause for excessive flywheel run out:

    A. Loose flywheel retaining nut
    B. Wrong size key in keyway
    C. Improperly installed flywheel
    D. Electrical components not installed behind flywheel

13. (Fill in the blanks) Crankcase breathers are usually _____ _____ or _____ _____ types.

CHAPTER 6

SERVICING THE 2-CYCLE ENGINE

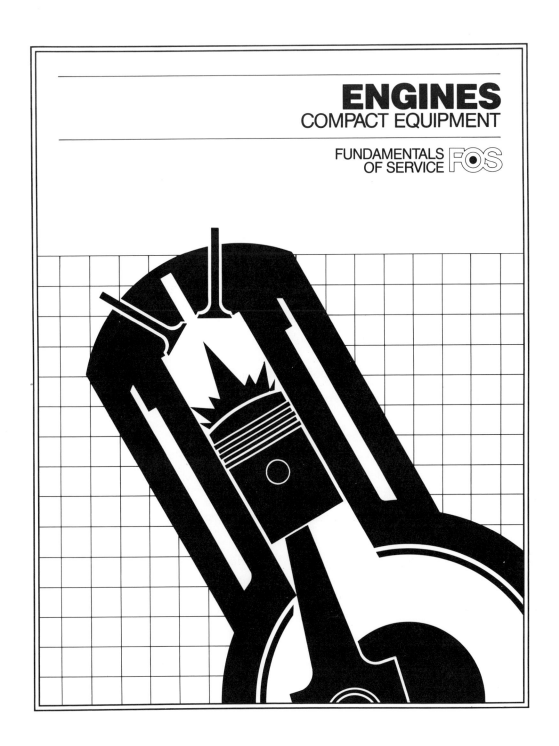

ENGINES
COMPACT EQUIPMENT

FUNDAMENTALS
OF SERVICE FOS

SKILLS AND KNOWLEDGE

This chapter contains basic information that will help you gain the necessary subject knowledge required of a service technician. With application of this knowledge and hands-on practice, you should learn the following:

• How to remove, clean, inspect, and install the cylinder head

• The procedure for removing and inspecting the cylinder

• How to recondition the cylinder

• How to disassemble and clean pistons and connecting rods

• What to look for when inspecting the piston and connecting rod components for damage

• Procedures for assembling the piston and connecting rod

• How to disassemble the flywheel, crankcase, and crankshaft

• What to look for when examining the flywheel and crankshaft for excessive wear

• Procedure for reassembling the crankshaft, crankcase, and flywheel

• How to check crankshaft end play

• How to measure crankshaft run out

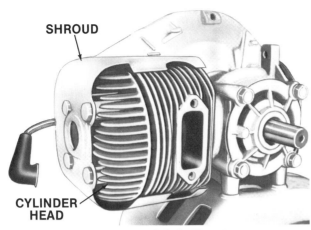

Fig. 1 — Remove Cylinder Head Bolts

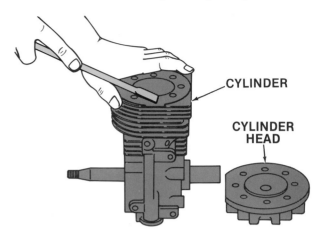

Fig. 3 — Clean Old Gasket from Cylinder Head and Cylinder

When service or repair is required, each engine component requires special treatment. The following pages describe disassembly, cleaning, inspection, and reassembly of each component of a 2-cycle engine. Engine diagnosis and troubleshooting is covered in Chapter 10.

A new safety feature on some 2-cycle engines used on lawn mowers is an engine brake. Because of the various brake designs, coverage of specific service procedures is not included in this text. The brake must be inspected and serviced periodically. Refer to the technical manual for service procedures.

CYLINDER HEAD

DISASSEMBLY

Loosen the head bolts and disassemble the cylinder head from the cylinder block (Fig. 1). If baffles or shrouds are present, remove them also. Inspect the cylinder head gasket prior to and during disassembly for signs of gasket burning. Never pry on the mating surfaces of cylinder head and cylinder block to loosen a stuck cylinder head. Instead, tap the cylinder head lightly with a soft hammer.

Many 2-cycle engines do not have removable cylinder heads. They are part of the cylinder and are removed when the cylinder is removed (Fig. 2).

CLEANING

Scrape all traces of old gasket material from mating surfaces of the cylinder head and the cylinder (Fig. 3). Remove carbon deposits from the combustion chamber. To avoid gouging, especially with aluminum cylinder heads and cylinder blocks, use scrapers that will not damage the cylinder head.

Fig. 2 — Engine with Removable Cylinder Head (Left), Nonremovable Cylinder Head (Right)

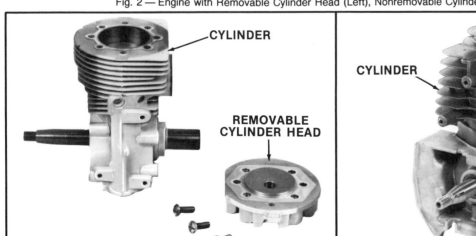

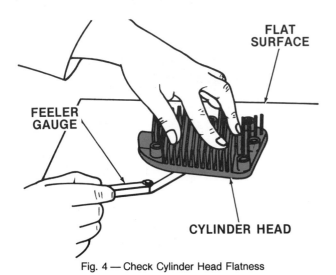

Fig. 4 — Check Cylinder Head Flatness

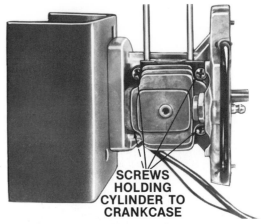

Fig. 6 — Remove Screws to Split Crankcase

INSPECTION

Check the cylinder head for cracks or other damage. Use a magnetic particle test or fluorescent dye to be sure. A severely damaged cylinder head can sometimes be repaired by a qualified specialty shop but will normally require replacement.

Check the cylinder head for warping. Hold the cylinder head down solidly on a flat surface and use the feeler gauge to determine flatness (Fig. 4).

ASSEMBLY

Always use a new cylinder head gasket and install it properly. Some cylinder head gaskets are easily reversed; the top of the gasket is usually marked "TOP" to avoid errors. Follow the manufacturer's recommendations on sealant (if any) to use with the cylinder head gasket.

Fig. 5 — Cylinder Head Bolt Tightening Sequence

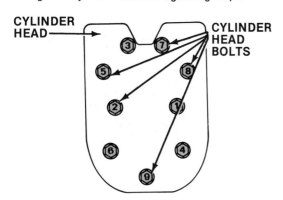

The technical manual illustrates the proper sequence for tightening cylinder head bolts (Fig. 5) in progressive stages until all bolts reach full torque as given in the manual. After the cylinder head is in place, install the spark plug.

Some manufacturers suggest that after a short run-in period the cylinder head bolts be retightened.

CYLINDER

DISASSEMBLY

On 2-cycle **engines with removable cylinders**, remove the screws that hold the cylinder to the crankcase (Fig. 6). Tap the side of the cylinder with a plastic hammer, if necessary, to loosen the cylinder from the crankcase. Remove the cylinder from over the piston (Fig. 7) and remove the cylinder base gasket.

On some 2-cycle engines, the **cylinder is an integral part of the crankcase** (Fig. 8). The piston and connecting rod must first be removed before the following cylinder inspection can be done. Piston and connecting rod disassembly for this type of engine is discussed in a following section.

INSPECTION

If the cylinder does not show extensive gouging, determine the amount and type of wear. Use an inside micrometer (Fig. 9) or telescoping bore gauge to measure the diameter of the cylinder at the top and bottom of the piston ring travel.

For **engines with a one-piece cylinder head and cylinder**, this measurement must be made by inserting

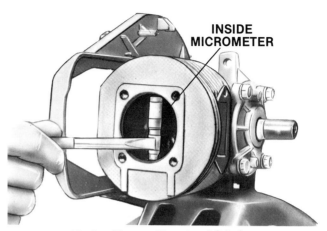

Fig. 9 — Measure Diameter of Cylinder

Measure the diameter twice at each location — once parallel to the axis of the crankshaft and again at right angles to the crankshaft axis. These two measurements show the amount the cylinder is out-of-round at that location. Check at both top and bottom to establish the amount of cylinder taper.

Taper and out-of-round measurements must not exceed the manufacturer's specifications. Bore diameter measurements over the maximum require that the cylinder be rebored or replaced.

For 2-cycle engines found on equipment such as snow throwers, chain saws, weed trimmers, and lawn mowers, it is usually more costly and time consuming to rebore a cylinder than to replace it. Most manufacturers, therefore, recommend replacement and do not specify oversized rings for rebored cylinders.

If wear is within the tolerance limits some manufacturers recommend that cylinders can be **honed or deglazed.** On engines with a one-piece cylinder head and cylinder, honing is possible but not practical. Chrome-plated cylinders should not be honed.

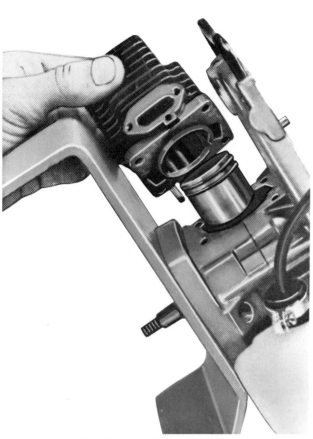

Fig. 7 — Removing Cylinder from Engine

the gauge into the bottom of the cylinder bore after the cylinder has been removed (Fig. 10).

Fig. 8 — One-piece Cylinder and Crankcase

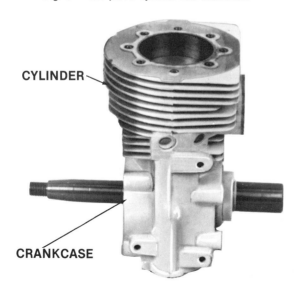

CYLINDER

CRANKCASE

Fig. 10 — Measure Cylinder Diameter on Engine with Nonremovable Cylinder Head

INSIDE MICROMETER

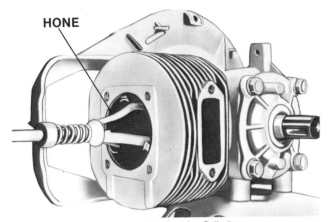

HONE

Fig. 11 — Honing the Cylinder

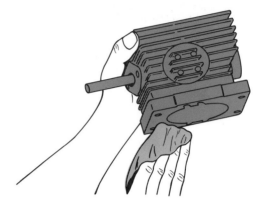

Fig. 13 — Clean Mating Surfaces of Cylinder and Crankcase

Honing Or Deglazing

A cylinder hone (Fig. 11) is a tool for removing high spots from the cylinder and for establishing a surface finish conducive to proper ring seating. Select the hone grit size which will produce the recommended surface on the cylinder wall. Hones from by various manufacturers will have different lubricant requirements. Follow the instructions which come with the hone.

Fit the hone to the narrowest part of the bore so it drags freely up and down the bore when it isn't rotating. After the hone is fitted, start it rotating and move it up and down the full length of the bore. Maintain a constant up and down movement. High spots will show up as areas of increased drag.

Keep the hone well lubricated, if it is a wet hone. Wire brush the cutting surfaces occasionally to prevent them from filling up with metal particles.

When the high spots have been reduced, make a final pass up and down the bore using the same constant speed strokes to impart an even, cross-hatched finish to the entire bore (Fig. 12). This finish will hold oil in the bore surface during piston ring break in. Avoid overhoning. Remove only as much metal as necessary to provide a smooth, cross hatched bore.

After honing, recheck the bore diameter as previously described. Check the readings against the manufacturer's tolerances. Proper honing should not change the diameter of the bore enough to make it out of tolerance. If it does, the cylinder was probably worn near its limit before honing.

Wipe reconditioned cylinder walls free of all abrasives, wash with hot water and detergent, and wipe clean. Wash and wipe until a clean white rag can be wiped firmly in the bore and show no discoloration. When the cylinder is clean, oil it with clean, light engine oil.

ASSEMBLY

When installing a new cylinder, clean all contact surfaces on both the crankcase and cylinder (Fig. 13). Install a new cylinder base gasket and make certain any cut-outs for transfer ports are properly aligned. Tighten the screws that hold the cylinder to the crankcase to the recommended torque.

Cylinders which do not have removable cylinder heads must have the piston and connecting rod installed before installing the cylinder. Piston and connecting rod assembly is covered in a following section.

Fig. 12 — Cross-hatched Finish after Honing

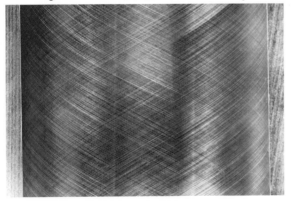

Fig. 14 — Removing Piston Pin

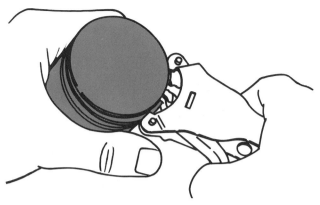

Fig. 16 — Removing Piston Rings with Ring Expander

PISTON AND CONNECTING ROD

DISASSEMBLY

The procedure for removing the piston from the connecting rod is the same regardless of engine design. The procedure is:

1. Remove the retaining rings from each side of the piston pin.

2. Push the piston pin out of the piston (Fig. 14). If the piston pin cannot be removed by hand, use a piston pin tool (Fig. 15).

3. Remove the piston and then the needle bearings, spacers, or bushing from the connecting rod eye.

After the piston has been removed, remove the piston rings with a ring expander (Fig. 16). Do not break the piston rings to remove them because that could damage the ring groove.

Connecting rod disassembly procedures vary depending on engine design. In fact, on many 2-cycle engines the connecting rod cannot be removed from the crankshaft (Fig. 17).

On these engines, with nonremovable connecting rods, the piston must be removed before the crankcase and crankshaft can be disassembled. Crankcase disassembly will be covered in a following section.

On an engine with a **removable connecting rod**, first remove the carburetor and reed valves or the bottom cover of the crankcase to expose the connecting rod capscrews (Fig. 18).

When the connecting rod capscrews are accessible, proceed as follows:

1. Place a clean cloth under the engine. The crankpin bearings usually are needle bearings which could fall out when the connecting rod cap is removed.

Fig. 17 — Connecting Rod Not Removable from Crankshaft

Fig. 15 — Piston Pin Tool

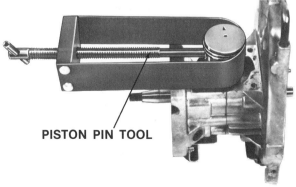

PISTON PIN TOOL

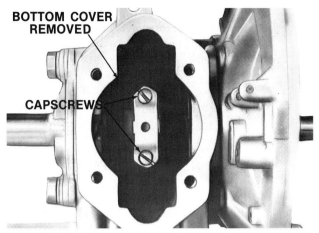

Fig. 18 — Connecting Rod Cap and Capscrews

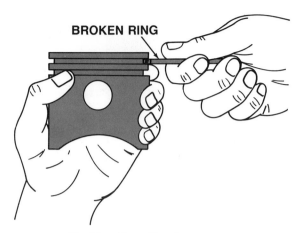

Fig. 20 — Clean Ring Grooves

2. Remove the connecting rod capscrews and the connecting rod cap. When disassembling connecting rod caps, make sure that each cap remains with its own connecting rod. Mark both the cap and the connecting rod so that the cap goes back on in the same position as it was removed. If the connecting rod doesn't already have aligning marks use a prick punch to put a light mark on each half of the connecting rod, adjacent to each other, but on opposite sides of the split point (Fig. 19). Realign the marks during assembly.

3. Push the connecting rod and piston out through the cylinder bore if the piston has not already been removed.

4. Remove needle bearings (if used) from the crankpin, connecting rod, and connecting rod cap. Count the number of bearings to make sure all are located. The

technical manual indicates the number of needle bearings used.

5. Remove the piston pin as described earlier.

CLEANING

Clean the head of the piston with fine sandpaper to remove carbon deposits. Use a piece from an unusable piston ring (Fig. 20) or ring groove cleaning tool to remove deposits from the ring groove.

Clean the connecting rod, cap, and bearings with a cleaning solvent.

INSPECTION

Inspect the cleaned piston for cracks, burned or overheated areas, and scoring or scuffing. A scored piston may be caused by lean fuel-air mixtures, dirty cylinder fins, or too little oil in the fuel-oil mixture. Replace cracked, heat damaged, or severely scored pistons.

Fig. 19 — Marking Connecting Rod and Cap

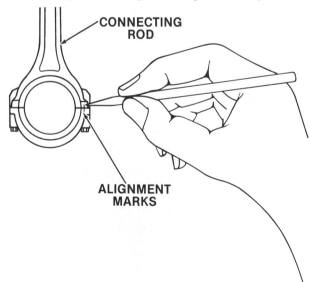

Fig. 21 — Measure Ring End Gap

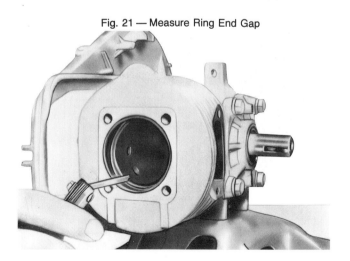

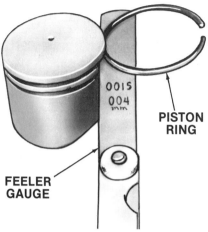

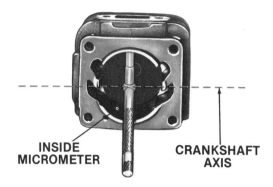

Fig. 24 — Measure Cylinder Bore at Right Angle to Crankshaft Axis

Fig. 22 — Measure Ring Groove Wear

Check piston rings for wear by inserting them into the cylinder about 1/2 inch (1.3 cm) from the top of the cylinder. Use a feeler gauge to check the ring end gap (Fig. 21). Replace rings that are not within manufacturer's specifications. If new rings are used, check end gap in the same manner.

Check the ring grooves for wear by inserting the edge of a new ring into the groove and checking the

remaining gap with a feeler gauge (Fig. 22). If the clearance measured is not within the manufacturer's specification, install a new piston.

Check piston diameter for excessive wear by measuring at the top and bottom of the skirt at right angles to the pin bore (Fig. 23). These measurements may indicate a lightly elliptical shape. This characteristic is normal since different areas of the piston expand at different rates as the piston warms up. Then check the measurement against those in the manual. Reject the piston if the wear is excessive.

Fig. 23 — Measure Piston Skirt

The piston may be within wear limits but still be too small for the bore if the bore has been reconditioned. Check the diameter of the cylinder at right angles to the crankshaft axis at the lower end of the bore (Fig. 24). The difference between this diameter and the readings taken from the lower end of the piston skirt will give the total piston clearance.

Check these measurements against the specifications to determine if piston replacement is required. If

Fig. 25 — Measure Piston Clearance in Cylinder

Fig. 26 — Measure Piston Pin Bore in Piston

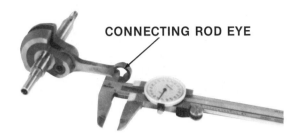

Fig. 28 — Measure Connecting Rod Eye

suitable bore gauges and micrometers are not available, insert thin feeler stock between the piston and cylinder wall at the bottom of the cylinder (Fig. 25) and use a light scale to measure the force needed to withdraw the stock. Many manufacturers specify piston clearance in this way.

Examine the piston pin bore in the piston for scratching or excessive wear. Measure the bore in two dimen-

sions, at right angles to each other, for size and out-of-round condition (Fig. 26). Measure the piston pin and inspect in the same way (Fig. 27).

Compare the measurements and replace the piston or piston pin if either is not within the manufacturer's specified limit.

Also measure the bore of the connecting rod eye to determine if it is within specification (Fig. 28). If not, install a new connecting rod.

If crankpin (connecting rod) bearings are serviceable needle bearings, check bearing condition and replace any that are damaged. Check the condition of the bore of the connecting rod head and cap. If it is rough or corroded, install a new connecting rod and cap.

Visually inspect the connecting rod for cracks, bends, or twists. An irregular wear pattern on the piston pin bearings and the crankpin bearing surface will generally indicate a bent or twisted connecting rod. The irregular wear pattern is more difficult to detect on needle bearings, which are most common in 2-cycle engines, than on plain bearings.

Fig. 27 — Measure Piston Pin

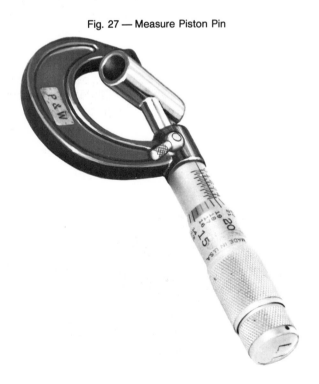

Fig. 29 — Check Side Clearance between Connecting Rod and Crankshaft Throw

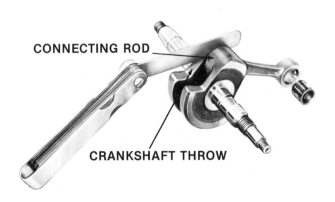

CONNECTING ROD

CRANKSHAFT THROW

Fig. 30 — Install Piston Pin

Use a feeler gauge to check side clearance between the connecting rod and the cheeks of the crankshaft throw (Fig. 29).

If the connecting rod is bent or twisted, or if the side clearance is not within the manufacturer's specification, replace the connecting rod.

ASSEMBLY

Use the following procedure to assemble and install the piston and connecting rod assembly.

1. Install the piston pin into one side of the piston pin bore (Fig. 30). Place a new retaining ring in the opposite side of the bore.

2. Lubricate (with 2-cycle engine oil) and install the needle bearing into the connecting rod eye.

Some engines require that the piston be heated before the piston pin is installed to make it easier for the pin

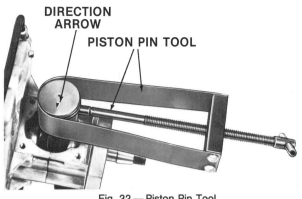

Fig. 32 — Piston Pin Tool

to fit into the piston pin bore. On these engines, completely submerge the piston in a container filled with SAE 30 motor oil. Use a heavy wire and bend a hook on each end to suspend the piston in the oil (Fig. 31). Do not let the piston touch the bottom of the container because the piston may be warped when the container of oil is heated.

Place the container of oil on a hot plate and gradually heat it until the oil begins to smoke. Remove the piston and proceed.

3. Place the piston over the connecting rod. Make sure the piston is installed in the correct direction. An arrow or mark is usually found on the head of the piston to indicate the correct direction (Fig. 32). The technical manual will tell the direction the arrow should point.

4. Align the piston pin with the needle bearing and push the piston pin through the bearing. If the piston pin cannot be pushed through by hand, use a piston pin tool (Fig. 32).

5. Install the other new retaining ring.

6. Install piston rings onto piston using a ring expander (Fig. 33). Use 2-cycle engine oil to lubricate the piston rings. Position the end gaps of the piston rings around

Fig. 31 — Suspend Piston in Container of Oil

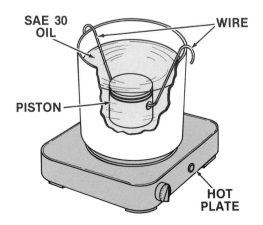

Fig. 33 — Install Piston Rings with Ring Expander

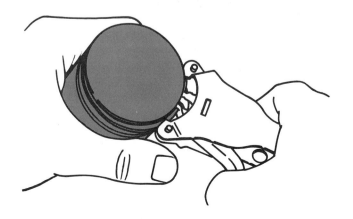

Fig. 34 — Locating Pins in Ring Groove

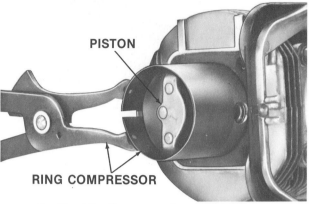

Fig. 36 — Ring Compressor Around Piston Rings

locating pins in the ring grooves (Fig. 34). This is necessary so the gap will not line up with an intake or exhaust port which could cause the ring to break.

If locating pins are not present in the ring grooves position the ring end gaps on opposite sides of the piston so they will not line up with intake or exhaust ports.

If piston rings have chamfered edges, install them as shown in Fig. 35.

Once the piston is assembled on the connecting rod, it must be installed in the cylinder.

On engines with a removable cylinder head and a removable connecting rod, proceed as follows.

1. Lubricate the piston and rings with 2-cycle engine oil.

2. Compress the piston rings with a ring compressor (Fig. 36).

3. Carefully place the connecting rod into the cylinder bore from the top. Make sure the connecting rod does not scratch the bore and that the piston exhaust side is on the same side as the cylinder exhaust ports.

4. Install the needle bearings around the crankpin of the crankshaft to which the connecting rod will be attached (Fig. 37). Apply a light grease to the crankpin so the needle bearings will not fall out during assembly.

5. Install bearing guides, if used, on the crankpin and bearing liners in the bore of the connecting rod and cap.

6. Push the piston into the cylinder so that the bearing liner in the connecting rod head fits against the needle bearings (Fig. 38).

7. Assemble connecting rod cap to the connecting rod around the crankpin (Fig. 39). Tighten the connecting rod capscrews to the specified torque.

Fig. 37 — Installing Needle Bearings Around Crankpin

Fig. 35 — Position of Chamfered Piston Rings

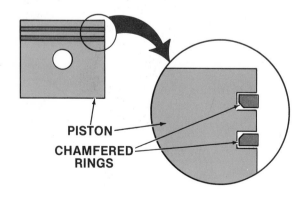

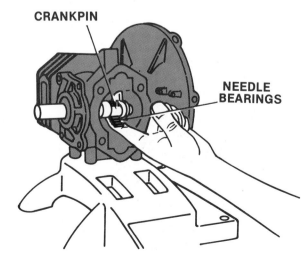

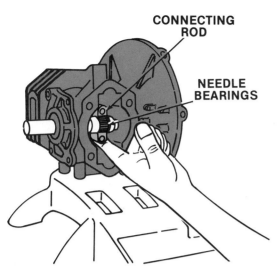

Fig. 38 — Bearing Liner Fits Against Needle Bearings

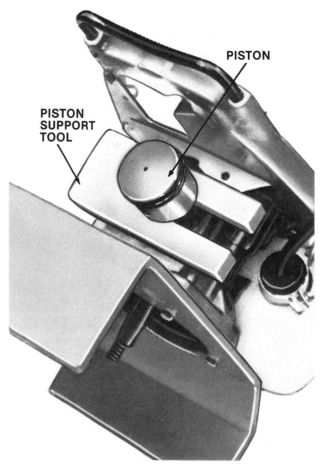

Fig. 40 — Piston Support Tool

On **engines with cylinder heads that are not removable**, the piston must be installed from the bottom of the cylinder. The procedure to use is:

1. Lubricate the piston and rings with 2-cycle engine oil.

2. Support the piston with a support tool (Fig. 40).

3. Install the cylinder over the piston. Rock the cylinder back and forth gently while sliding it down over the piston rings (Fig. 41). Compress the rings with fingers or with a ring compressor.

IMPORTANT: Do not rotate cylinder while installing piston rings. Rotating could cause the rings to go out of alignment with the locating pins.

4. Install cylinder on crankcase, as described earlier, and install spark plug.

Fig. 41 — Installing Cylinder over Piston

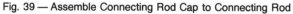

Fig. 39 — Assemble Connecting Rod Cap to Connecting Rod

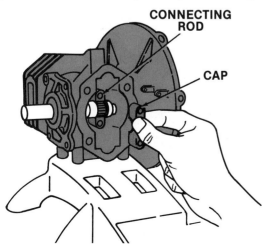

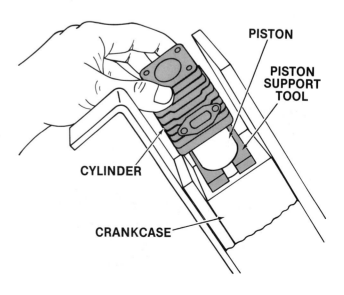

Fig. 42 — Tool to Hold Flywheel While Removing Nut

FLYWHEEL, CRANKCASE, AND CRANKSHAFT

DISASSEMBLY

Many 2-cycle engines contain a **split crankcase** which must be disassembled to remove the crankshaft. To disassemble:

1. Remove the flywheel locking nut with a wrench or socket while holding the flywheel with a filter wrench or special tool (Fig. 42).

FLYWHEEL PULLER
FLYWHEEL

Fig. 43 — Flywheel Puller

Fig. 44 — Remove Clutch Assembly (if used)

CLUTCH ASSEMBLY

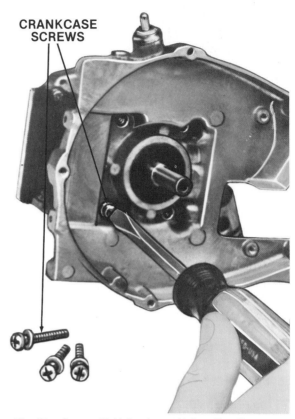

CRANKCASE SCREWS

Fig. 45 — Screws Hold Crankcase Halves Together

2. Remove the flywheel by using a flywheel puller (Fig. 43). Never pry on the flywheel to try to remove it. Prying can bend the flywheel or crankshaft or can damage components mounted behind the flywheel.

3. Remove electrical components and clutch assembly (if used) (Fig. 44).

4. Remove end keys from both ends of crankshaft

5. Remove the screws that hold the crankcase halves together (Fig. 45).

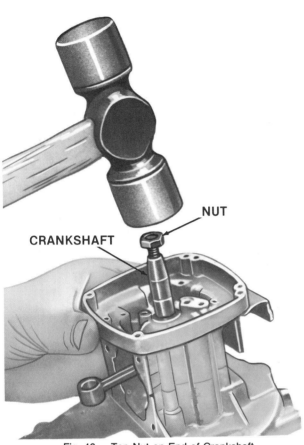

Fig. 46 — Tap Nut on End of Crankshaft

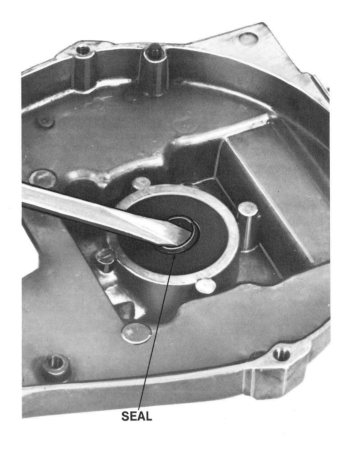

Fig. 48 — Pry Out Seals with Screwdriver

6. Place the flywheel locking nut back on the end of the crankshaft to protect the threads. While holding the crankcase, use a plastic hammer to tap the nut on the end of the crankshaft from the ignition side (Fig. 46). This should separate the crankcase halves.

Another method of separating crankcase halves involves holding one half of the crankcase and tapping the other half with a plastic hammer to separate the halves (Fig. 47).

Crankshaft bearings and oil seals generally remain in the crankcase halves when the engine is disassembled. To remove seals, pry them out with a screwdriver (Fig. 48). To remove the bearing, use a bearing driving tool and tap it with a hammer (Fig. 49).

Fig. 47 — Tap Crankcase Half to Separate

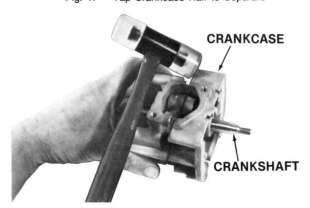

Fig. 49 — Bearing Driving Tool

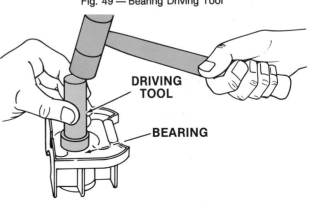

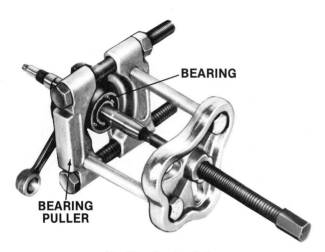

Fig. 50 — Bearing Puller

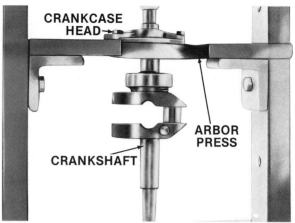

Fig. 52 — Press Out Bearing and Crankshaft

If the bearings remain on the crankshaft after disassembly, use a bearing puller to remove them (Fig. 50).

If the engine contains a **one-piece crankcase**:

1. Remove the screws that hold the crankcase head assembly to the engine (Fig. 51).

2. Tap the end of the crankshaft, from the ignition side, to drive the crankshaft out of the crankcase (Fig. 51).

3. Place crankcase head in an arbor press and apply light but steady pressure to press out bearing and crankshaft (Fig. 52). Also press the oil seal from the crankcase head. Other types of oil seals may be removed with a screwdriver as shown in Fig. 48.

INSPECTION

Observe crankcase for excessive internal wear. Measure crankshaft main bearing journals (Fig. 53) and crankpin journals for excessive wear. Replace the crankshaft if the measurement does not fall within the tolerance specified by the manufacturer.

Fig. 53 — Measure Main Bearing Journals

Clean the flywheel and examine it for cracks and broken cooling fins (if so equipped). Replace the flywheel if these conditions exist.

Check the starter ring gear (if equipped) for broken or badly damaged teeth. The ring gear can be replaced

Fig. 54 — Removing Damaged Starter Ring Gear

Fig. 51 — Drive Crankshaft Out of Crankcase

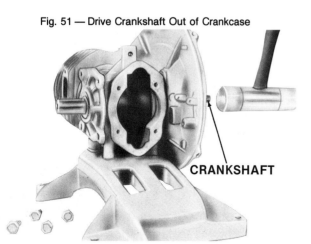

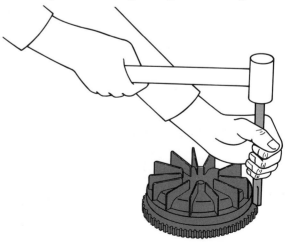

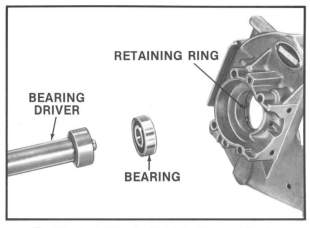

Fig. 55 — Install Bearing Retaining Ring and Bearing

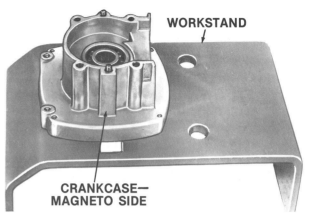

Fig. 57 — Workstand to Assemble Crankcase

on some flywheels by cutting off the old ring gear (Fig. 54). A new ring gear can be installed by heating it and dropping it squarely into position on the flywheel. Take care not to overheat. Assure that the gear bevels face the proper direction and that the gear is fully seated before it cools. Most 2-cycle engines don't have replaceable ring gears and the flywheel must be replaced.

ASSEMBLY

To install new oil seals and crankshaft bearings proceed as follows:

1. Install the retaining ring into its groove in the crankcase bore (Fig. 55) in each crankcase half.

2. Drive the bearings into the bores with the proper tool until the bearings are seated against the retaining rings (Fig. 55).

3. Lubricate the seal lips and outer surfaces of the seal cases with high-temperature grease prior to installation.

4. Install the new oil seals by tapping them into the crankcase bore until flush (Fig. 56).

Remaining assembly includes assembling the crankshaft and the crankcase halves. If a workstand is available, use it to secure the magneto side of the crankcase during assembly (Fig. 57). Assemble the components as follows:

1. If a thrust washer is used, install it on the magneto end of the crankshaft.

2. Insert the magneto end of the crankshaft into the magneto side of the crankcase (Fig. 58). Gently tap the opposite end of the crankshaft with a plastic hammer until it is seated.

NOTE: Position the threaded end of the crankshaft toward the ignition.

Fig. 56 — Install Oil Seal

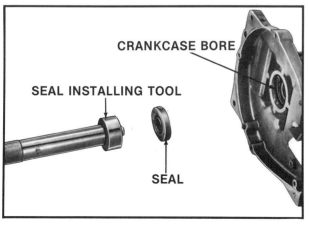

Fig. 58 — Install Crankshaft into Crankcase

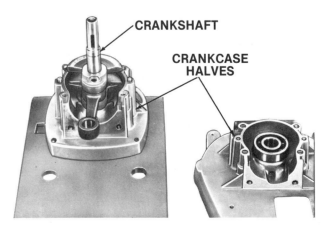

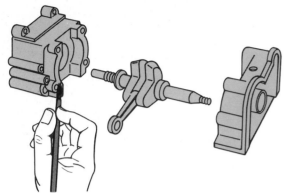

Fig. 59 — Applying Sealant to Mating Surface of Crankcase Half

3. Install a new gasket between the crankcase halves or use the recommended sealant if no gasket is used. (Fig. 59). Use the recommended sealant to coat both sides of the gasket.

4. Make sure the surfaces around the dowel pins are clean. If they are not, the crankcase halves will not mate properly.

5. Install another thrust washer, if used, on the opposite end of the crankshaft.

6. Install the other crankcase half by tapping it gently into place with a plastic hammer. Make sure the dowel pins are aligned with the holes.

7. Install the screws that hold the crankcase halves together (Fig. 60). Use a thread sealer if recommended in the technical manual. Alternately snug the screws to pull the crankcase halves together. Then tighten the screws to the recommended torque.

8. Check for crankshaft rotation. If the crankshaft does not turn easily, strike the ends of the crankshaft with a plastic mallet. If this does not correct the problem, disassemble and clean the crankcase and crankshaft.

Fig. 60 — Install Screws to Hold Crankcase Together

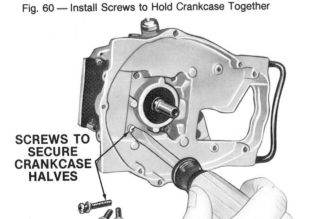

SCREWS TO SECURE CRANKCASE HALVES

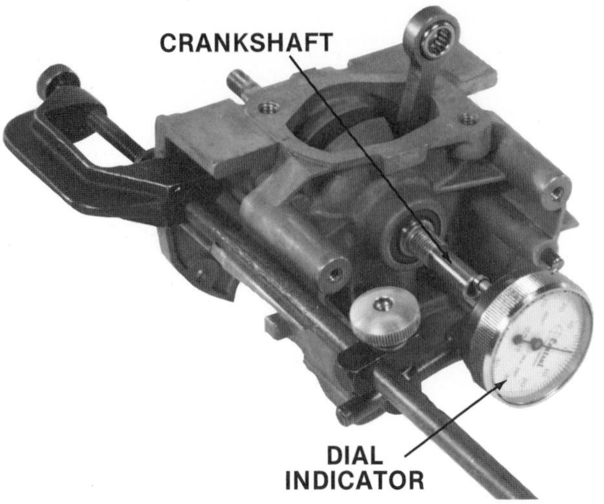

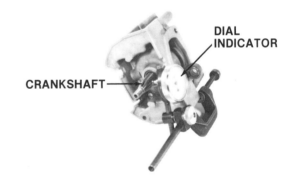

Fig. 61 — Check Crankshaft End Play

9. Use a dial indicator to check crankshaft end play (Fig. 61). If end play is not within the manufacturer's specification, replace thrust washers (if used), crankshaft, or crankcase.

10. Measure crankshaft run out by attaching a dial indicator to rest against the crankshaft (Fig. 62). Rotate the crankshaft by hand and observe the reading on the dial indicator. If run out is not within manufacturer's specifications, install a new crankshaft.

DIAL INDICATOR

CRANKSHAFT

Fig. 62 — Checking Crankshaft Run out

Fig. 63 — Place Flywheel over Key

FLYWHEEL KEY

FLYWHEEL

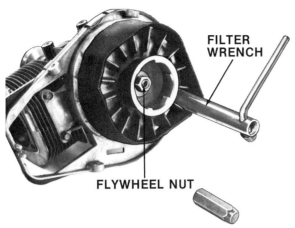

Fig. 64 — Tighten Flywheel Nut

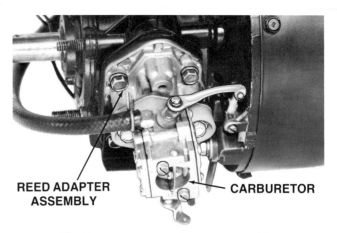

Fig. 66 — Removing Reed Adapter Assembly

After the crankcase has been assembled, use the following procedure to install the flywheel:

1. Install clutches (if used) and electrical components.

NOTE: Clutches may be installed either before or after the flywheel depending on design.

2. Install the flywheel key in the crankshaft.

3. Carefully place the flywheel over the key (Fig. 63). The flywheel hub may be made of aluminum and may be easily damaged.

4. Apply thread sealant, if recommended in the technical manual, to threads on the end of the crankshaft.

5. Install the washer and flywheel nut and hand tighten.

6. Hold the flywheel with a filter wrench or other special tool and tighten the flywheel nut to the specified torque (Fig. 64).

After all engine components have been assembled, attach other exterior components (such as fuel tank, carburetor, and muffler) according to instructions in the technical manual.

REED VALVES

On engines that contain reed valves, the valves and adapter assembly must be removed and inspected periodically.

DISASSEMBLY AND INSPECTION

Before disassembling reed valves check the reeds for leakage. Remove the air cleaner from the carburetor intake and hold a piece of paper about one inch (25 mm) from the intake (Fig. 65). If spots of fuel begin to appear on the paper, the reeds are leaking.

Remove the carburetor mounting screws (Fig. 66) and remove the carburetor from the engine. Remove the screws holding the reed adapter assembly (Fig. 66) and remove the adapter.

Fig. 65 — Check for Leakage of Reeds

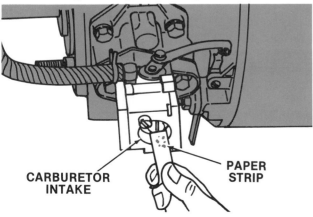

Fig. 67 — Various Arrangements for Reed Valves

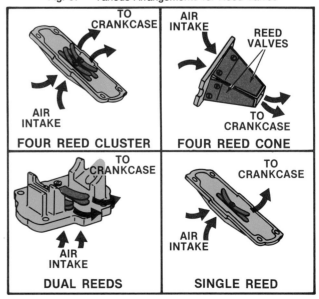

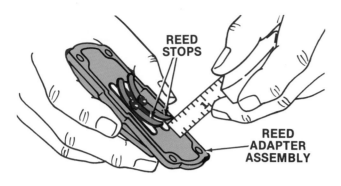

Fig. 68 — Check for Bent or Distorted Reeds

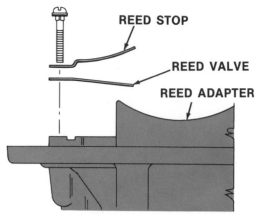

Fig. 69 — Installing Reeds

The reed valve assembly may have different arrangements for the reed valves (Fig. 67) but in all the arrangements, the reed must seat against a reed plate. When inspecting reeds, check this seat for the proper seal. There should be no clearance between the reed and reed plate.

Also, inspect for cracked or bent reeds. If the assembly contains a reed stop (to prevent the reed for opening too far), check it for bends or distortion (Fig. 68). Replace cracked or bent reeds, but reed stops may be bent back into position.

Examine the bottom of the reed plate for excessive warpage. Leakage in this area will cause low crankcase pumping pressure and will result in too little fuel-air mixture for the engine.

ASSEMBLY

The reed should be slightly curved to provide a good seal with the reed plate. Install reeds so that the curve (pressure side) is down against the reed plate (Fig. 69). After assembling the components, use a new gasket and install the adapter assembly on the engine.

SUMMARY

When servicing a 2-cycle, each component requires careful cleaning, inspection, reconditioning (if needed), and assembly.

The cylinder head may or may not be removable. If it can be removed, inspect the cylinder head and clean gasket material from the mating surfaces of the cylinder head and cylinder. Assemble the cylinder head with a new cylinder head gasket.

The cylinder also may or may not be removable. Most engines with nonremovable cylinders have removable cylinder heads. The removable cylinder head is required for these engines so the piston can be removed.

Some cylinders may be reconditioned with a hone.

Procedures for piston and connecting rod disassembly vary depending on whether the cylinder is removable or the cylinder and crankcase are in one piece. The piston pin, piston rings, connecting rod needle bearing, connecting rod cap, and crankpin bearings may be disassembled for inspection and replacement. Measurements may be made on most components to determine if they are within the manufacturer's specifications.

Many 2-cycle engines contain a split crankcase which must be disassembled to remove the crankshaft. The flywheel must first be removed. The crankcase may then be disassembled and the crankshaft bearings and oil seals removed for inspection or replacement.

CHAPTER 6 REVIEW

1. (True or false) Some 2-cycle engines contain a cylinder head that is cast in one piece with the cylinder.

2. (Fill in the blanks) Cylinders may either be _____ or they may be _____ with the crankcase.

3. (True or false) The connecting rod on some 2-cycle engines cannot be removed from the crankshaft.

4. Describe the procedure to use to check the end gap of new or used piston rings.

5. (True or false) Most 2-cycle engines use needle bearings as crankpin bearings.

6. Why must the piston in some 2-cycle engines be heated in a container of oil before the piston pin is inserted?

7. Why are locating pins often used in the ring grooves of 2-cycle engine pistons?

8. (Fill in the blanks) Two crankcase designs commonly found on 2-cycle engines are:_____ and _____ - _____.

9. (Fill in the blanks) Reeds may be checked for leakage by holding a piece of paper in front of the _____ side of the _____.

CHAPTER 7
SERVICING THE GASOLINE
AND DIESEL FUEL SYSTEMS

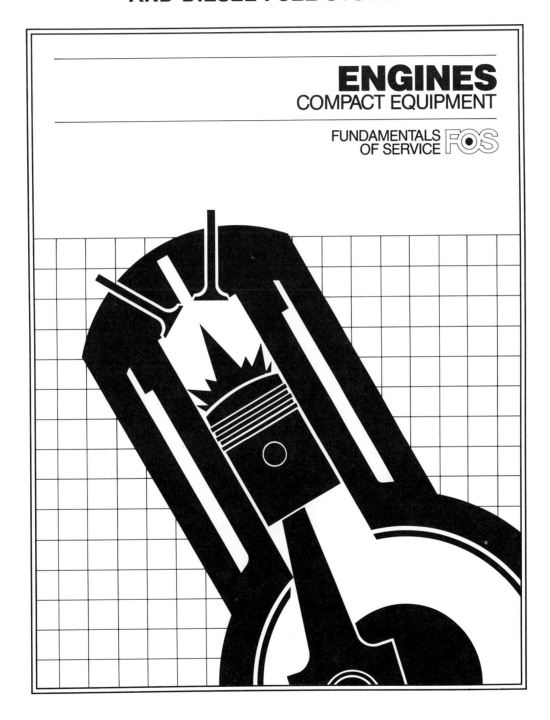

ENGINES
COMPACT EQUIPMENT

FUNDAMENTALS
OF SERVICE FOS

SKILLS AND KNOWLEDGE

This chapter contains basic information that will help you gain the necessary subject knowledge required of a service technician. With application of this knowledge and hands-on practice, you should learn the following:

Gasoline Fuel System

• How to remove a carburetor from an engine

• The steps involved in disassembling the carburetor

• Special disassembly procedures for float-type, suction-lift type, and diaphragm-type carburetors.

• How to clean, inspect, and repair each type of carburetor

• How to assemble and install carburetors

• What types of adjustments can be made on carburetors

• The procedures for disassembling, inspecting, repairing and adjusting a governor

• How to service the fuel filter

Diesel Fuel System

• How to remove and install a fuel-injection pump

• The procedure for timing an injection pump

• How to bleed air from a diesel fuel system

• How to remove and test an injection nozzle

• The steps required to disassemble a nozzle

• How to clean and inspect injection nozzle parts

• The service procedures required of diesel fuel filters

• How to disassemble and inspect a precombustion chamber

• How to remove, install, and adjust the governor

• What tests are required before completely disassembling a turbocharger

• The procedures for disassembling, cleaning, and inspecting turbocharger parts

Components Common To Both Systems

• Service procedures for dry-element and oiled-element type air cleaners

• How to service the fuel pump

• How to service and repair the fuel tank and fuel lines

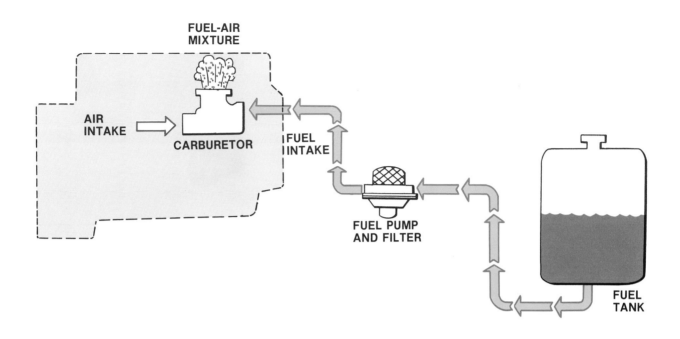

Fig. 1 — Gasoline Fuel System

INTRODUCTION

Gasoline and diesel fuel systems vary considerably in components and operation as discussed in Chapter 4. Therefore, the servicing requirements and adjustments are different for each type of fuel system.

The first part of this chapter will cover service and adjustment procedures for gasoline engines and the second part of the chapter will cover diesel fuel system service.

Several components are common to both gasoline and diesel fuel systems. Service procedures for these components will be covered at the end of the chapter after the discussion on diesel fuel system service. These components are:

- **Air cleaner**
- **Fuel tank**
- **Fuel pump**
- **Fuel lines**

GASOLINE FUEL SYSTEM SERVICE

Components of the gasoline fuel system that require servicing are (Fig. 1):

- **Carburetor**
- **Governor**
- **Fuel pump**
- **Fuel filter**
- **Fuel tank**
- **Fuel lines**
- **Air cleaner**

Carburetors, governors, and fuel filters will be covered in this section. The remaining components will be covered at the end of the section COMPONENTS COMMON TO BOTH SYSTEMS.

CARBURETORS

As discussed in Chapter 4, the types of carburetors most common on compact equipment engines are:

- **Float type — adjustable and nonadjustable**
- **Suction-lift type — vacuum- and pulsating-lift types**
- **Diaphragm type**

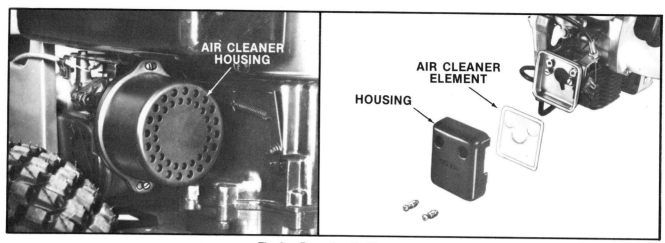

Fig. 2 — Removing Air Cleaner

Exploded views, which show the parts of each type of carburetor, are printed in the Appendix of this text.

Since different types of carburetors contain many parts common to all types, the discussion here will generally apply to all types. If a service procedure is unique to a specific type of carburetor, the type of carburetor being discussed will be noted.

The service procedures covered in this chapter include:

- **Removal from engine**
- **Disassembly**
- **Inspection and repair**
- **Reassembly and installation**
- **Adjustments**

Two basic points to remember when servicing the carburetor are:

- **Service** the carburetor at these times:

1. After engine valve grinding or engine overhaul.

2. Every year or at the beginning of each season on seasonal machines. At this time, clean the carburetor, check all components, and replace gaskets and seals.

- **Adjust** the carburetor at these times:

1. During engine tune up

2. After a major engine overhaul

3. Whenever the carburetor has been removed for service

Removal From Engine

Before removing the carburetor from the engine, close the fuel shutoff valve. Next remove the air cleaner and its housing (Fig. 2) and disconnect governor, throttle, and choke linkages (Fig. 3). Remember the holes to which the linkages are assembled so they can be reassembled in the same position.

Disconnect the fuel line and breather hose (Fig. 3) (if equipped). Remove the carburetor from the engine, intake manifold, or reed adapter (2-cycle engines with reed valves).

Read the engine technical manual for specific procedures for removing the carburetor from a given engine.

Fig. 3 — Disconnect Various Linkages and Hoses

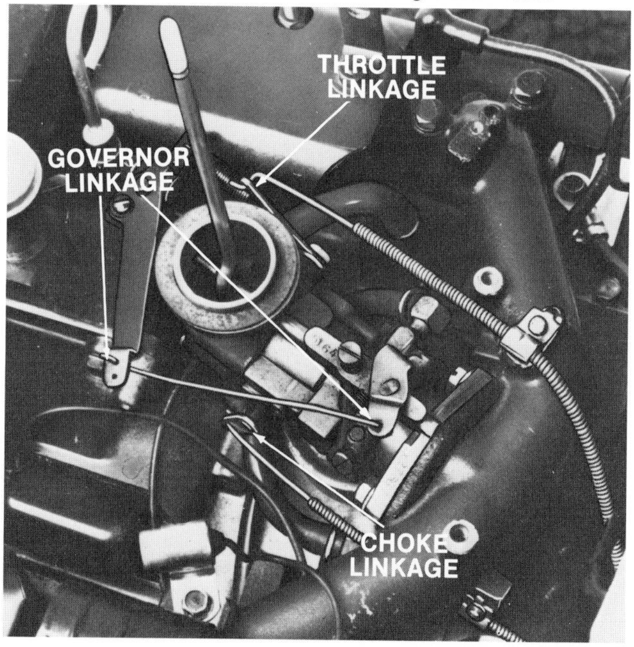

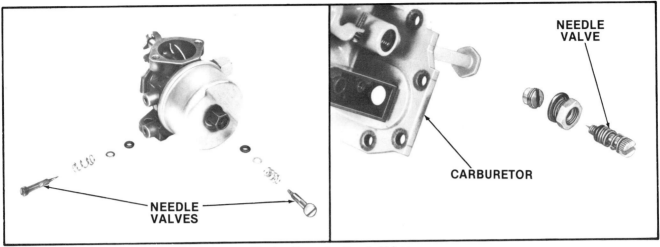

Fig. 4 — Removing Needle Valves

Disassembly

Refer to the Appendix of this text for exploded views of various types of carburetors.

Important: Clean the outside of the carburetor before removing it from the engine.

To disassemble the carburetor:

1. Remove the needle valve(s) (if applicable) for idle mixture, idle speed, and high-speed mixture (Fig. 4).

2. Remove a screw and washer to remove the choke valve (Fig. 5).

3. Lift the choke lever and spring from the carburetor (Fig. 5).

4. Remove a screw and washer to remove the throttle valve (Fig. 6).

5. Lift the throttle lever and spring from the carburetor (Fig. 6).

Other parts to remove during disassembly depends on the type of carburetor.

FLOAT-TYPE CARBURETOR

For further disassembly of the float-type carburetor:

1. Remove nut, washer, float bowl, and gasket (Fig. 7).

2. Remove the pin (not shown) from the hinge and remove the float from the carburetor (Fig. 8).

3. Slide the float valve and tip off the tab (Fig. 8).

Fig. 5 — Removing Choke Valve and Lever

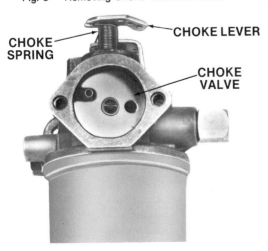

Fig. 6 — Removing Throttle Valve and Lever

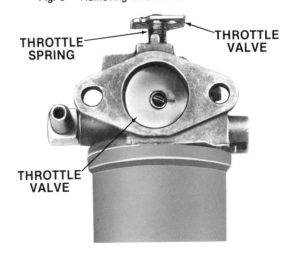

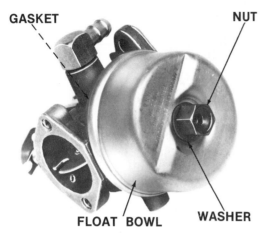

Fig. 7 — Removing Float Bowl

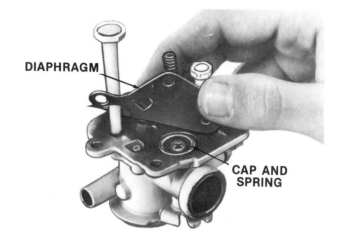

Fig. 10 — Removing Spring and Diaphragm

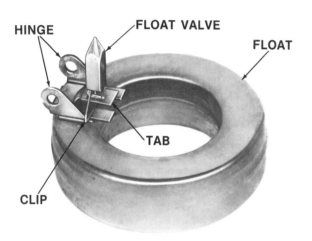

Fig. 8 — Removing the Float and Float Valve

4. Remove the float valve seat. To remove the seat it may be necessary to first place a few drops of oil on the seat. Then place a compressed air hose nozzle against the inlet fitting (Fig. 9) and use a short blast of air to blow the seat out of the carburetor body.

CAUTION: Compressed air can blow the seat out of the carburetor body at a high rate of speed. Wear safety glasses and direct the seat away from your body when using compressed air.

SUCTION-LIFT CARBURETOR

For further disassembly of the suction-lift carburetors:

1. Lift the diaphragm and remove the cap and spring (Fig. 10) for pulsating type.

2. Remove the cover and gasket from the automatic choke (Fig. 11).

Fig. 9 — Removing Float Valve Seat

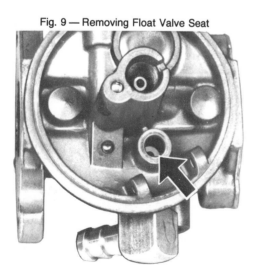

Fig. 11 — Removing the Diaphragm

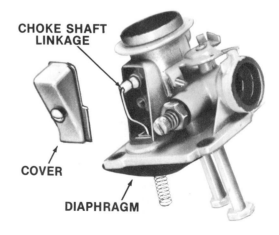

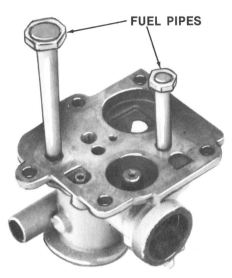

Fig. 12 — Removing Fuel Pipe(s)

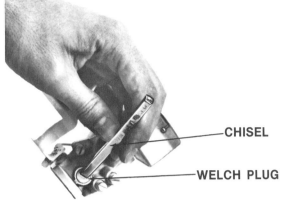

Fig. 14 — Removing Welch Plug

3. Disconnect the choke shaft linkage and remove the diaphragm (Fig. 11).

4. Use a socket wrench to turn the fuel pipes counterclockwise out of the carburetor (Fig. 12).

NOTE: The vacuum suction-lift carburetor will have only one fuel pipe. See the discussion on carburetor operation on pages 4-16 through 4-18.

DIAPHRAGM-TYPE CARBURETOR

For further disassembly of the diaphragm-type carburetor:

1. Remove the diaphragm cover and the diaphragm (Fig. 13).

2. Remove the welch plug (a cover for the idle fuel chamber) by driving a small chisel or punch into the

center of the plug (Fig. 14). The dent in the plug should release the plug's grip on the carburetor body and the plug should fall out. If it does not, use the chisel to pry the plug out of position.

Inspection and Repair

Repair kits are available for most carburetors. The kit contains parts most often needed for carburetor repair (Fig. 15). The kit is recommended for use if the carburetor is completely disassembled for service.

The first step in carburetor inspection is to clean all the parts. Place all metallic parts in a suitable basket and immerse the basket in a container of carburetor cleaning solution.

Some carburetors contain a fuel inlet seat fitting which is metal with a neoprene seat. Do not soak the seat in the solution and do not remove the seat. Refer to the engine technical manual for other parts that require special cleaning consideration.

Important: Never clean holes or passages with small drill bits or wire. A slight enlargement or burring of these

Fig. 13 — Removing Diaphragm

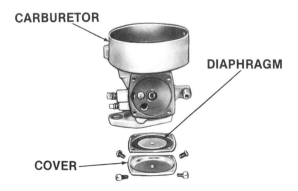

Fig. 15 — Carburetor Repair Kit

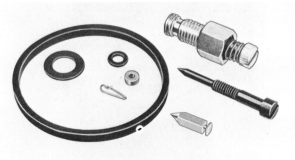

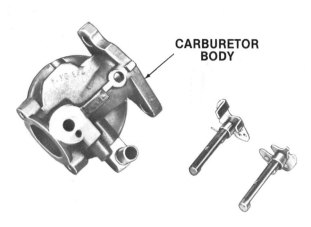

CARBURETOR BODY

Fig. 16 — Inspecting Carburetor Body

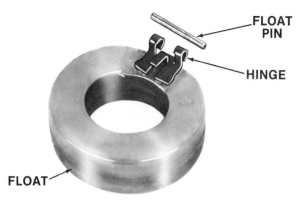

FLOAT PIN

HINGE

FLOAT

Fig. 18 — Inspecting the Float

holes will change the performance of the carburetor. Soaking in solvent is the only cleaning method that should be used.

Leave the parts in the solution for one to two hours. Then transfer the parts to a clean solvent. After rinsing in the solvent, dry each part with compressed air, making sure all holes are open and free of dirt.

Never use rags or paper to dry the parts. These materials may leave lint which may plug passages and affect carburetor operation. Clean carburetor air passages with compressed air.

Important: Never use compressed air to clean a completely assembled float-type carburetor. To do so may cause the float to collapse.

After cleaning a carburetor body that is made of aluminum rinse it in hot water to neutralize any corrosive action the cleaner may have on the aluminum. Check the engine technical manual to see if this procedure is recommended for a specific carburetor.

After cleaning the parts, inspect the parts which will be reused for damage. Inspect the carburetor body (Fig. 16) for cracks or damage.

Inspect the throttle shaft, choke shaft, and shaft bearing surfaces in the carburetor body (Fig. 16). Excessive clearance will give poor engine performance and may allow dirt particles to enter the engine. Replace these parts as necessary.

Inspect mixture needles for wear at the seating surfaces (Fig. 17). Check springs and O-rings on mixture needles and replace if they are damaged. Inspect the needle seats for scoring.

On **float-type carburetors**, inspect the float for dents, leaks, or worn hinge bearing surfaces (Fig. 18). Also inspect the float pin for wear.

Inspect the float valve (Fig. 19) for wear. The valve and seat come as a set. If either is damaged, replace both.

Fig. 19 — Inspecting Float Valve

Fig. 17 — Check Condition of Mixture Needles

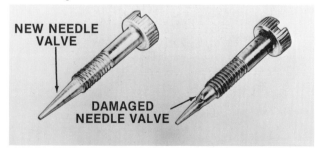

NEW NEEDLE VALVE

DAMAGED NEEDLE VALVE

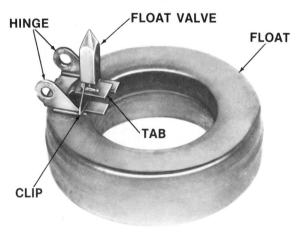

HINGE

FLOAT VALVE

FLOAT

TAB

CLIP

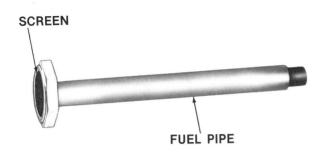

Fig. 20 — Inspecting Fuel Pipes

Inspect the float bowl gasket and float bowl drain assembly. Replace the gasket if damaged. The gasket on the bowl drain assembly is also replaceable if leakage is noticed.

On **suction-lift type carburetors** inspect the fuel pipes (Fig. 20) for clogs or damage. The screen must be clean and the check ball free. Replace the fuel pipe if it cannot be cleaned or is damaged.

Inspect the diaphragm (Fig. 21) for wear or punctures. Remove the spring and measure the length. Replace the diaphragm and spring if they are damaged.

Since the suction-lift carburetor sits on top of the fuel tank, the fuel tank must also be inspected during carburetor servicing. Place a straightedge across the machined flat surface of the fuel tank (Fig. 22). Use a feeler gauge to measure clearance between the fuel tank surface at the points highlighted in the inset and the straightedge.

If clearance is too great, fuel leakage between the fuel tank and diaphragm may allow excess fuel in the carburetor. In this case, install a repair kit.

Fig. 21 — Inspecting Diaphragm

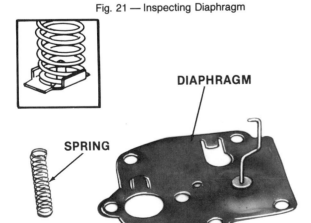

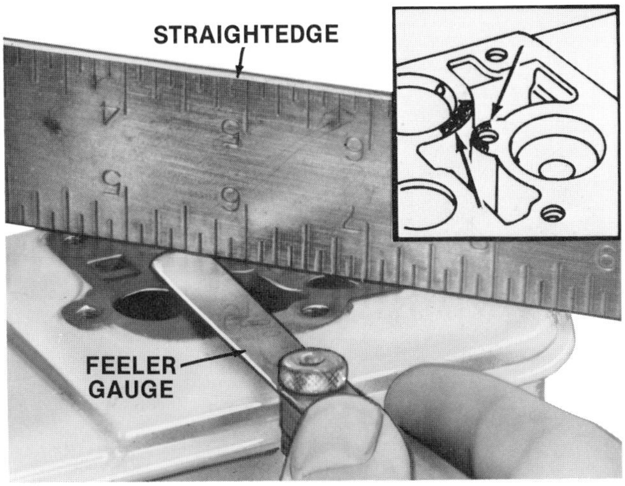

SPRING

DIAPHRAGM

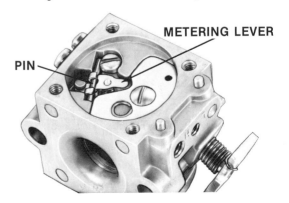

Fig. 22 — Measuring Fuel Tank Top Surface

NOTE: Symptoms of a warped fuel tank are:

- *The engine running at idle speed with the needle valve completely closed*

- *Fuel in the diaphragm spring chamber of the fuel tank*

On **diaphragm-type carburetors** check the diaphragm for cracks, punctures, distortion, or deterioration. Replace a damaged diaphragm or gasket.

Some diaphragm-type carburetors have a metering lever to meter fuel (Fig. 23). Check the lever and replace it if is bent, twisted, or worn. The lever should move freely and fit well with the metering lever pin.

Reassembly and Installation

To assemble the carburetor, place the throttle lever and spring in the carburetor and install the throttle valve (Fig. 24). The throttle valve may be marked to indicate the proper position for the valve.

Fig. 23 — Carburetor with Metering Lever

METERING LEVER

PIN

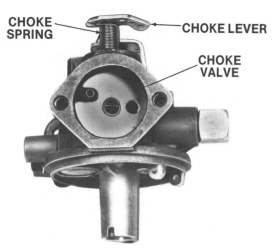

Fig. 24 — Installing Throttle Valve

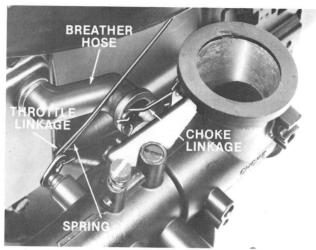

Fig. 26 — Installing Fuel Line and Breather Hose

Next, place the choke lever and spring in the carburetor and install the choke valve (Fig. 25).

Install the mixture needles and tighten them until they are finger tight. Then back the needles out about one full turn. Check the operator's manual or technical manual for the exact initial setting.

To install the carburetor on the engine, place the mounting screws in place but do not tighten. Install throttle, choke, and governor linkages (if equipped) and install fuel line and breather hose (Fig. 26).

Then tighten the mounting screws and adjust the carburetor. Adjustment procedures will be covered in a following section.

FLOAT-TYPE CARBURETOR

On float-type carburetors, before the carburetor is installed, install the float valve seat (Fig. 27) by inserting the grooved side into the carburetor body first. Lightly oil the cavity in the carburetor body and then use a flat face punch to push the seat into position.

Fig. 27 — Installing Float Valve Seat

Fig. 25 — Installing Choke Valve

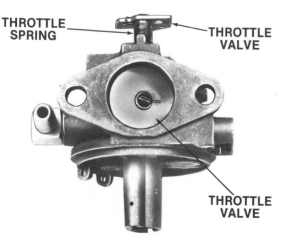

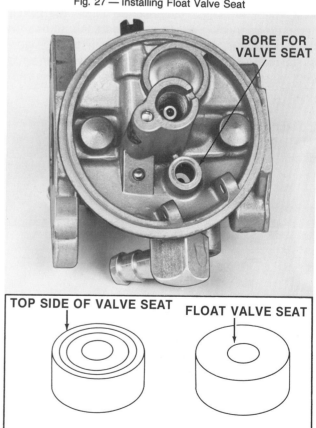

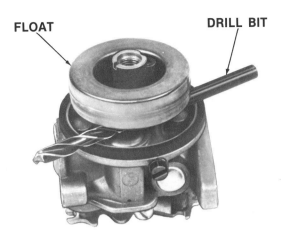

Fig. 28 — Installing Float Bowl and Gasket

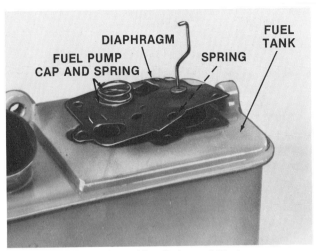

Fig. 30 — Assembling Spring and Diaphragm

Assemble the float valve and the clip on the tab (see Fig. 19). Place the float in the carburetor body and install the float pin through the hinge.

After assembling the float, invert the carburetor to check the float adjustment. Use a ruler or a drill bit of the size recommended in the technical manual to get an accurate measurement (Fig. 28). If adjustment is necessary, remove the float and bend the tab on the float until the correct setting is obtained.

Install the float bowl and gasket and tighten the retaining nut.

Install the carburetor on the engine.

SUCTION-LIFT CARBURETOR

Additional assembly required before installing the suction-lift carburetor is to first install the fuel pipe(s) (Fig. 29). Screw the pipe into the carburetor body and tighten with a wrench or socket.

If the carburetor is equipped with a diaphragm (for the automatic choke or for a pulsating-type carburetor), first connect the spring to the diaphragm (inset, Fig. 30). Place the diaphragm on the top of the fuel tank making sure the spring is in the spring pocket.

Next place the fuel pump cap and spring on the diaphragm and set the carburetor on the fuel tank (Fig. 31). Align the mounting screw holes in the carburetor, diaphragm, and fuel tank. Connect all linkages, make sure all components are in place, and then install mounting screws.

DIAPHRAGM-TYPE CARBURETOR

Additional assembly, before installation, of the diaphragm-type carburetor includes installing the welch plug if it was removed for cleaning. The welch plug is a cover for the idle fuel chamber.

Fig. 31 — Installing Carburetor on Fuel Tank

Fig. 29 — Installing Fuel Pipe(s)

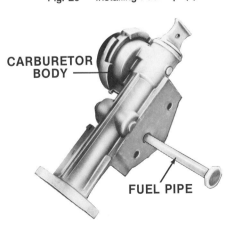

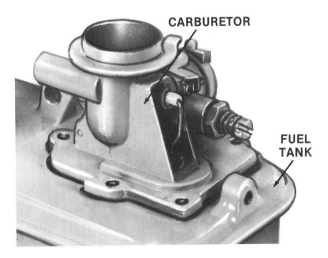

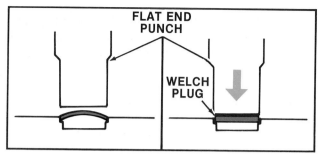

Fig. 32 — Installing Welch Plug

Before installing the welch plug, clean its receptacle in the carburetor. Then install the plug in the receptacle with the raised part of the plug up (Fig. 32). Then flatten the plug so it will remain in place with a flat-end punch that is equal to or greater in size than the plug.

Next, install the gasket, diaphragm, and cover (Fig. 33). Check the technical manual to make sure the diaphragm is installed with the correct side toward the carburetor.

Install the carburetor on the engine.

Adjustments

Each carburetor (if adjustable at all) is adjusted in a slightly different manner. The procedure is to adjust the idle speed needle, idle mixture needle, and the high-speed mixture needle (Fig. 34) until the engine runs smoothly. The following is a typical example of how to adjust the carburetor.

Before making any adjustments, obtain a vibration tachometer, if the engine does not have an electric tachometer, to check engine speed (Fig. 35).

Fig. 33 — Installing Gasket, Diaphragm, and Cover

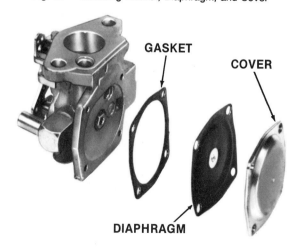

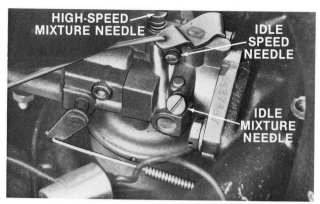

Fig. 34 — Adjust Speed and Mixture Needles

• To adjust a carburetor high speed operation:

1. Start the engine and allow it to warm up.

2. Set engine speed at the recommended speed to adjust the high-speed mixture needle.

3. Slowly turn the high-speed mixture needle closed until the engine starts to lose speed (lean mixture). Note the position of the needle.

4. Slowly turn the needle the opposite direction, past the point of smoothest operation until the engine just begins to run rough (rich mixture).

5. Turn the needle closed until it is positioned half way between the lean mixture and rich mixture positions.

• To adjust a carburetor for idle operation:

1. Set the throttle control lever to the SLOW or IDLE position. If this does not give the recommended idle speed, proceed to step 2.

2. Turn the idle-speed screw clockwise to increase idle speed or counterclockwise to decrease idle speed.

Fig. 35 — Vibration Tachometer

Fig. 36 — Removing Governor Assembly

3. After obtaining the correct idle speed, slowly turn the idle mixture needle closed until the engine just begins to run rough.

4. Turn the idle mixture needle the opposite direction until the smoothest idle mixture is obtained.

GOVERNOR

Governor service procedures will vary depending on the type of governor used. The following discussion will cover disassembly, inspection, assembly, and adjustment of a typical flyweight and air-vane type governor. Refer to the technical manual for specific procedures to use on a given governor.

Disassembly

Disassembly of the **flyweight governor** was covered on page 5-12. It is disassembled with the camshaft. The

Fig. 37 — Disconnecting Governor Springs and Linkages

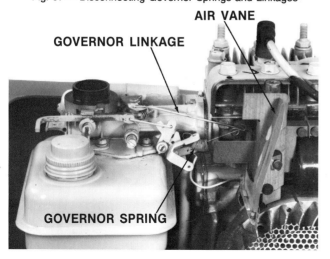

disassembly procedure involves disconnecting the linkages, removing the timing gear cover, and removing the governor assembly (Fig. 36).

To remove the **air-vane governor**, remove the blower shroud and air cleaner. Disconnect the governor springs and linkages (Fig. 37). Remove the screws holding the air vane and remove the vane.

When disassembling the governor linkages, note the holes in the levers from which the governor links and springs were removed. They must be reinstalled in the same holes.

Inspection and Repair

For the **flyweight governor**, inspect the governor gear for broken or worn teeth and the plunger and flyweights for binding or damage. Inspect other components (such as levers, springs, and shafts) for damage and replace as necessary.

On **air-vane governors** inspect the air vane, springs, links, and levers for wear or damage and replace parts as necessary.

On both types of governors, check to see that the throttle is open when the engine is stopped. If the throttle is not open, the governor is out of adjustment or a part of the governor assembly is bent or damaged.

Assembly and Installation

Install the governor assembly back on the engine and install the shroud (air-vane type). On flyweight governors, make sure the governor gear meshes properly with the crankshaft or camshaft gear. Connect all springs and linkages in the same position that they were before disassembly.

Fig. 38 — Adjusting Governor Lever

Adjustment

Before starting the engine to make any adjustments, first check to see that the throttle valve is open when the engine is stopped. If the throttle is not open, adjust the governor lever or arm until it is open (Fig. 38). Before attempting to adjust the governor, start the engine and allow it to warm up.

When the engine is warm, advance the throttle control lever to the wide open position and check engine speed with a tachometer. If engine **speed is higher or lower** than the top speed recommended by the manufacturer, first inspect the governor linkages to see if any are binding or bent. If they are not, the governor spring probably needs replacing.

If the **engine hunts or surges,** reposition the governor spring in another hole in the lever or arm. If this doesn't stop the hunting or surging, replace the spring.

FUEL FILTER

Fuel filters on compact equipment gasoline engines are often a simple, single-element filter (Fig. 39). The only service required is to remove the element periodically

Fig. 39 — Single-Element Fuel Filter

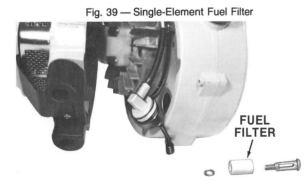

FUEL FILTER

and either wash it in a solvent or replace with a new filter. If the filter is washed in a solvent, let it dry completely before reinstalling it in the fuel system.

DIESEL FUEL SYSTEM SERVICE

Components of the diesel fuel system that require servicing are:

- **Fuel injection pump**
- **Fuel injection nozzles**
- **Precombustion chamber**
- **Governor**
- **Air cleaner**
- **Fuel tank**
- **Fuel lines**
- **Fuel pump**
- **Fuel filter**
- **Turbocharger (if equipped)**

Injection pumps and nozzles, precombustion chamber, governors, fuel filters, and turbochargers will be covered in this section. Service procedures for all other components will be covered in the next section COMPONENTS COMMON TO BOTH SYSTEMS.

The procedure for bleeding air from the diesel fuel system will also be discussed later in this section.

FUEL INJECTION PUMP

Fuel injection pump service at a compact equipment service facility is usually limited to the following procedures:

- **Removal for replacement**
- **Installation**
- **Checking and adjusting timing**

For injection pump disassembly and inspection of internal components, it is recommended that pumps be sent to diesel service centers which specialize in injection pump repair.

The fuel injection pump may be disassembled at the service department if the department is equipped with the proper tools and test stand. Only properly trained and qualified technicians should attempt this type of service procedure.

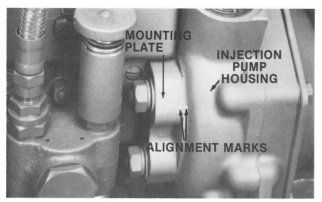

Fig. 40 — Alignment Marks on Injection Pump

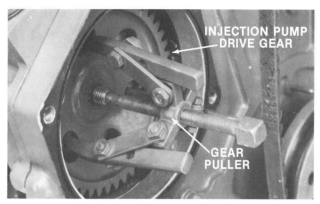

Fig. 42 — Removing Injection Pump Drive Gear

Removal

Before removing the fuel injection pump, clean the pump, lines, and the area around the pump with a cleaning solvent or a steam cleaner.

Important: Never steam clean or pour cold water on an injection pump while the pump is running or when it is warm. To do so may cause seizure of pump parts.

Before removing the pump, close the fuel shutoff valve. Then proceed as follows:

1. If alignment marks are not machined into the parts, place marks on the injection pump housing and its mounting plate (Fig. 40).

2. Disconnect the speed control linkage, the fuel supply line, and injection lines (Fig. 41). Plug all openings on the pump and fuel lines, preferably with plastic caps and plugs. Do not use cloth or a fibrous material.

3. Disconnect the governor spring and linkages. Do not distort or bend springs and links during disassembly.

Fig. 41 — Disconnect Linkages and Fuel Lines

4. On some pumps, it may be necessary to remove the injection pump drive gear with a gear puller (Fig. 42).

5. Remove the mounting bolts and then the injection pump. Some pumps are equipped with shims under the pump mounting flange. These shims are used to adjust injection pump timing. Note the number of shims removed so the same number can be installed later.

NOTE: Do not rotate the engine after the injection pump and gear are removed. Proper alignment during reassembly is essential for pump timing.

Installation

To install the fuel injection pump:

1. Use a new gasket between the injection pump housing and the mounting plate.

2. Fit the injection pump onto the engine and install the mounting bolts finger tight. If shims are used to adjust pump timing, install the shims under the pump mounting flange. Make sure all alignment marks are lined up if the pump being installed is the one that was removed from the engine. If a new pump is being installed, place it in as close to the same position as possible to the one that was removed.

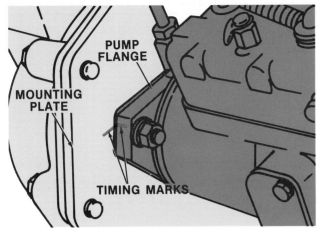

Fig. 43 — Lining Up Timing Marks

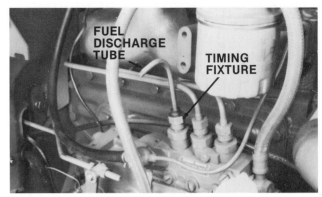

Fig. 45 — Installing Timing Fixture

3. On some fuel injection pumps, it is necessary to pivot the pump so timing marks will line up (Fig. 43). Pivot the pump before tightening the mounting bolts. On other pumps, tighten the mounting bolts to the torque specified in the technical manual.

4. Connect all linkages that were disassembled and attach all fuel lines and pipes that were removed.

Adjusting Injection Pump Timing

Injection pump timing adjustments depend on the type of pump being adjusted. Check the engine technical manual for specific adjustments for a given pump. The following discussion is a typical injection pump timing adjustment procedure.

1. Close the fuel shutoff valve.

2. Disconnect the injection line, at the injection pump leading to the No. 1 cylinder (Fig. 44). Remove the delivery valve from this line.

3. Install the recommended timing fixture on the delivery valve holder and install it on the pump (Fig. 45).

4. Use a screwdriver to turn the flywheel and crankshaft (Fig. 46) until the piston of the No. 1 cylinder approaches top dead center (TDC) of its compression stroke.

5. Place a small container under the fuel discharge tube of the timing fixture to collect fuel.

6. Open the fuel shutoff valve and continue to turn the flywheel.

NOTE: Since the delivery valve for the No. 1 cylinder was removed from the injection pump, fuel will flow from the timing fixture. The stream of fuel will be continuous up to a specific point of crankshaft rotation. The point at which the fuel flow just stops is the beginning of injection.

NOTE: If fuel does not stop flowing, even when the TDC mark has passed the pointer (Fig. 47), it is an indication that the No. 1 piston was on the exhaust stroke instead of the compression stroke.

Fig. 44 — Disconnecting Fuel Injection Line

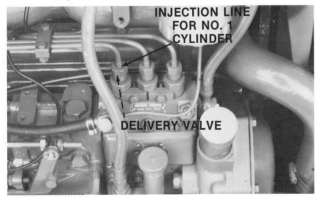

Fig. 46 — Turning the Flywheel

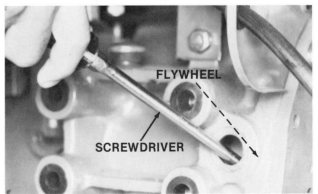

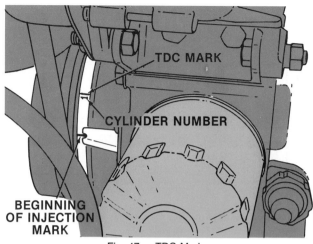

Fig. 47 — TDC Mark

7. Rotate the flywheel until fuel flow just stops. As the rotation approaches this point, fuel flow will change from a steady stream to drops. The drops will stop at the beginning of injection.

8. Check the position of the timing mark for beginning of injection (Fig. 47). If the timing is correct, they will be aligned with the pointer.

9. If the mark is not aligned with the pointer, loosen the pump mounting bolts and change the number of shims under the pump mounting flange.

10. Rotate the engine in the opposite direction about 1/4 of a revolution and then repeat steps 7-9.

11. Remove the timing fixture from the injection pump and install the delivery valve (see Fig. 44). The delivery valve must be lubricated with clean diesel fuel and must still be wet during installation.

12. Connect the fuel line with the delivery valve to the injection pump and tighten the connection to the recommended torque.

13. It will be necessary to bleed the fuel system after the pump has been adjusted.

BLEEDING AIR FROM DIESEL FUEL SYSTEMS

When the diesel fuel system has been opened for service (lines disconnected or filter removed), air can enter the system. If air is left in the lines, it may form an air lock which can prevent fuel from reaching or going through the injection pump. This could result in the engine misfiring, losing power, or not starting at all.

Be sure to bleed the system to remove this trapped air before operating the engine.

Here is a general guide to bleeding the system:

1. Fill the fuel tank with the proper diesel fuel.

2. Open the fuel shutoff valve.

3. Loosen the bleed screw at the fuel filter or water separator (Fig. 48). Pump the primer on the fuel pump (Fig. 49) until a solid stream of fuel (free of air bubbles) flows from the opening. (If the primer level will not pump fuel and no resistance is felt at upper end of stroke, turn the engine with the starter to change the position of fuel pump cam.) Tighten the screw. When bleeding is completed, be sure to leave the primer lever at lowest point of its stroke.

Fig. 48 — Loosening Bleed Screw

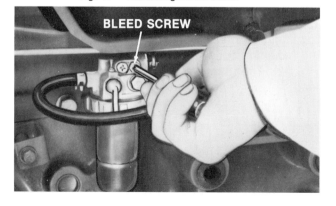

Fig. 49 — Pumping the Primer

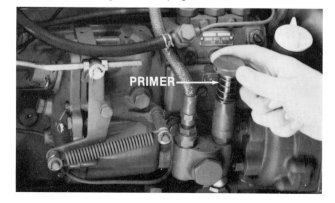

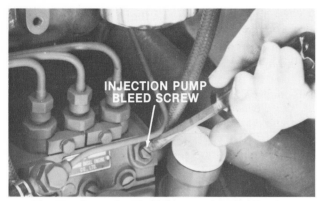

Fig. 50 — Loosening Bleed Screw on Injection Pump

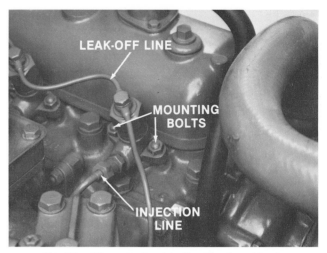

Fig. 52 — Disconnecting Injection Line from Nozzle

If the engine will not start, it may be necessary to bleed air from the injection pump. Loosen the bleed screw (Fig. 50) and operate the primer until fuel flow is free of bubbles. Tighten the bleed screw.

If the air lock is still present after bleeding air from the filter bleed screw, you will also have to bleed the injection lines.

4. Hold the injection nozzle fitting with a wrench and loosen the injection line nut on at least two injection nozzles (Fig. 51). Crank the engine until fuel without foam flows around the connectors. Tighten the line connections carefully. Avoid bending the line connection. Tighten the lines only until snug and free of leaks.

CAUTION: Loosen injection line connectors only one turn to avoid excessive spray.

CAUTION: Escaping fluid under pressure can penetrate the skin causing serious injury. While cranking the engine to bleed the system, do not allow anyone to stand near the loosened injection line nuts because fuel is being ejected under high pressure.

If any fluid is injected into the skin, it must be surgically removed within a few hours by a doctor familiar with this kind of injury or gangrene may result.

INJECTION NOZZLES

Fuel injection nozzle service consists of:

• **Removal**

• **Testing**

• **Disassembly**

• **Cleaning and inspection**

• **Assembly and installation**

Before any service is performed, clean each nozzle and the area around the nozzle.

Removal

To remove injection nozzles, first remove leak off lines and mounting bolts and disconnect the injection line (Fig. 52). Then remove the nozzle from the cylinder head.

Fig. 51 — Loosening Injection Line Nut

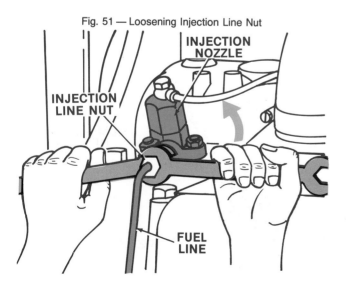

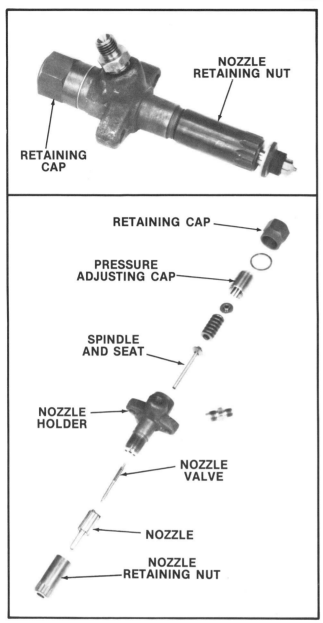

Fig. 53 — Removing Nozzle Retaining Nut

To disassemble the nozzle, place the nozzle body in a vise and use a wrench to remove the nozzle retaining nut (Fig. 53). Remove the nozzle and valve from the retaining nut.

Remove other parts from the nozzle body and retaining nut. Keep all mated parts together so they can be reassembled together.

Do not permit the lapped surfaces of some parts to contact any hard substance. This could damage the surface. Also, do not touch the nozzle valve unless hands are wet with diesel fuel.

Cleaning and Inspection

If new nozzles are used, they usually are packaged with a light coating of anti-corrosive grease. Wash the nozzle thoroughly in clean diesel fuel to remove grease.

Nozzles that will be reused may contain carbon deposits. Clean off the deposits by washing the nozzle in diesel fuel. If the parts contain hardened carbon or lacquer, use a brass wire brush to clean the deposits.

IMPORTANT: Never use a wire brush to clean nozzles. The wire can distort the nozzle.

After cleaning the exterior of the nozzle, inspect the lapped machined surface for nicks or scratches. Inspect the large part of the nozzle valve (Fig. 54) for scratches and scoring. Also check for a broken nozzle valve tip. Replace the nozzle valve if it is damaged.

Check the seats for both the nozzle valve and the nozzle. It may be necessary to use a magnifier to detect scratches on these seats. If scratches are present, replace the seat.

Disassembly

After the nozzle is tested it may be disassembled for cleaning and service. The nozzle contains many precision-made parts that are susceptible to damage by dirt and water. Disassemble nozzles in a spotlessly clean, dust-free area. Cover the workbench with clean paper before disassembling the parts.

As parts are disassembled, place them in a pan of clean diesel fuel and leave them until needed. Do not allow the parts to strike each other.

Fig. 54 — Inspecting Nozzle Valve

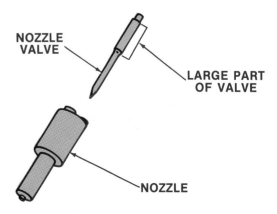

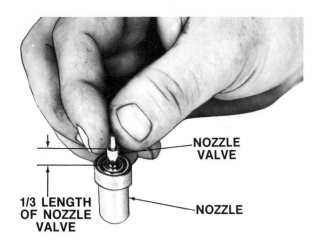

Fig. 55 — Checking Nozzle Valve Movement

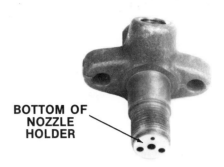

Fig. 57 — Inspect Bottom Surface of Nozzle Holder

NOTE: A damaged nozzle valve seat will allow fuel to drip from the nozzle. Damaged nozzle valve seats will usually be detected during the LEAKAGE TEST described earlier.

Check to see that the nozzle valve slides freely to its seat by using the following procedure.

1. Dip the nozzle valve in clean diesel fuel and insert the valve in the nozzle.

2. Hold the nozzle vertical and pull the valve out about 1/3 of the way (Fig. 55).

Fig. 56 — Cleaning the Nozzle Orifice

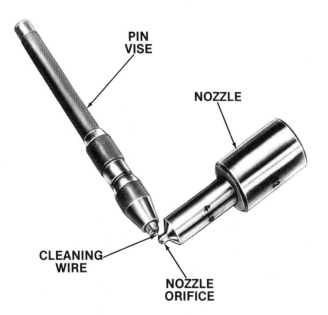

3. Release the valve. The weight of the valve should cause it to slide down to its seat.

Replace the nozzle assembly if the nozzle valve does not slide freely.

Next remove any carbon deposits from the spray orifices of the nozzle assembly. Use a cleaning wire, clamped in a pin vise, that is smaller than the size of the orifice (Fig. 56). Check the technical manual for orifice size. Insert the wire that protrudes from the vise into the orifice and rotate.

After cleaning the orifices, inspect the lapped machined surface on the bottom of the nozzle holder for nicks or scratches (Fig. 57). Replace the holder if it is damaged.

Blow out the fuel passages with compressed air and clean the threads.

Clean and inspect the threads of the nozzle retaining nut and the pressure adjusting cap (Fig. 58). Inspect upper seat and spindle seat for wear and replace as necessary. Also check for excessive wear in the area where the spindle contacts the nozzle valve stem.

Assembly and Installation

Before assembling the nozzle parts, make sure all parts are coated with clean diesel fuel. The technician's fingers should also be wet with diesel fuel.

Insert the nozzle valve into the nozzle while parts are immersed in clean fuel. Assemble the remaining parts in the order as shown in Figs. 53 and 58. Nozzles on different engines may contain some additional or different parts. See the technical manual for specific assembly instructions for each nozzle. Tighten all caps and nuts to the recommended torque.

Test the assembled nozzle as discussed earlier.

PRESSURE
ADJUSTING CAP

UPPER SEAT

SPINDLE SEAT

SPINDLE

NOZZLE VALVE

NOZZLE

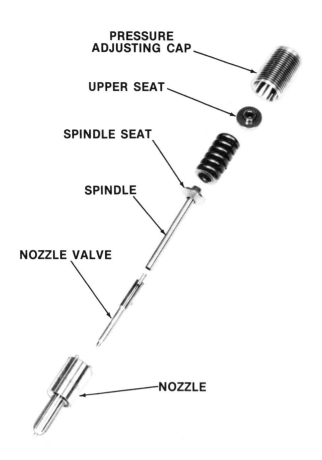

Fig. 58 — Inspect Nozzle Parts

After the nozzle is assembled and the exterior cleaned, insert the nozzle in its bore in the cylinder head. Connect all injection lines and then install the nozzle mounting bolts (Fig. 59). Next install leak-off lines to the nozzles and the pump and tighten all injection line connections to the recommended torque.

After assembly, bleed air from the fuel system as discussed earlier.

Fig. 59 — Installing Nozzle

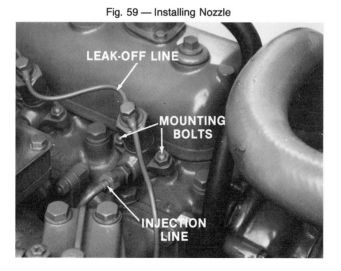

LEAK-OFF LINE

MOUNTING
BOLTS

INJECTION
LINE

WATER
SEPARATOR

BOWL

Fig. 60 — Fuel Shutoff

FUEL FILTER

Fuel filters on diesel engines are more complex than those found on gasoline engines. Diesel fuel filters must do a better job of filtering the fuel because diesel engine components are manufactured to closer tolerances than gasoline engine components. Therefore, the diesel engine is more susceptible to damage by contamination and water.

Some diesel fuel systems contain a separate water separator. Others have a water separation mechanism built into the fuel filter.

If the system contains a separate water separator, close the fuel shutoff lever before removing the bowl from the housing (Fig. 60). Next, remove the bowl and the filter screen from the housing (Fig. 61). Empty the bowl, and clean both parts in solvent.

Fig. 61 — Removing Bowl and Filter Screen

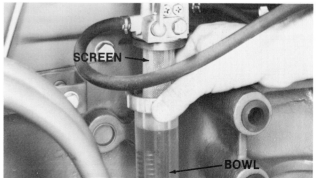

SCREEN

BOWL

Fig. 62 — Diesel Fuel Filter

Fig. 64 — Pump and Governor Chamber Covers

After cleaning, reinstall the screen and bowl and bleed air from the fuel system.

Fuel filter service consists of removing the old element and replacing it with a new one (Fig. 62). Never try to clean and reuse a fuel filter element.

PRECOMBUSTION CHAMBER

If a diesel engine contains a precombustion chamber, the chamber may be removed for service or replacement.

To remove the chamber, first remove the injection nozzle as described earlier. Then use a tool with a hook on one end, insert the tool in the precombustion chamber, and withdraw the chamber components (Fig. 63). If the chamber cannot be removed by this method, it may be necessary to remove the cylinder head and drive the chamber components out.

Fig. 63 — Precombustion Chamber

PACKING

PRECOMBUSTION CHAMBER

When disassembled, inspect components for wear. The heat insulation packing (Fig. 63) prevents much of the combustion heat from reaching the injection nozzle. If the packing is not in good condition, replace it.

GOVERNOR

In most diesel fuel systems, the governor is part of the fuel injection pump and is not serviceable. However, some governors are located outside the pump and may be removed and adjusted.

Removal

To remove the governor, first remove the cover plates from the pump chamber and the governor chamber (Fig. 64). Next, disconnect the governor spring and linkages to the injection pump.

NOTE: Do not drop parts during disassembly. They may fall through openings into the crankcase. To avoid this, place a shop towel over the opening to the crankcase during disassembly.

Assemble governor components in the reverse order that they were removed except leave the governor spring disconnected until the following adjustments are made.

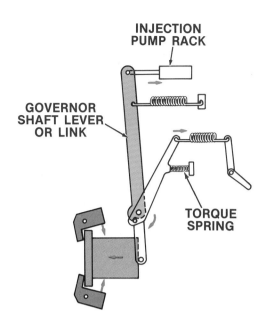

Fig. 65 — Adjusting Governor Arm

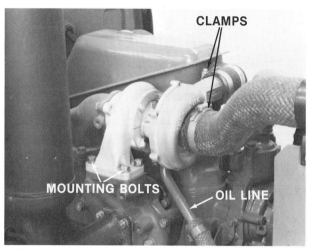

Fig. 67 — Disconnect Oil Line and Bolts from Turbocharger

To adjust this type of governor:

1. Move the governor arm toward the torque spring until it just touches the spring (Fig. 65).

2. Move the governor link one way or the other until reference marks on the control rack and pump housing align (Fig. 66).

Fig. 66 — Aligning Pump Reference Marks

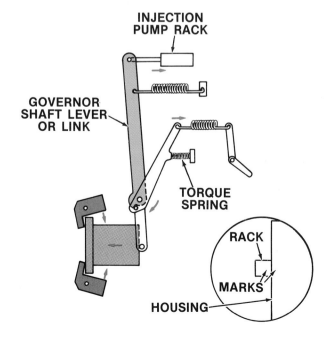

3. Hold the components in this position with a screwdriver and tighten the nut (A) in the linkage.

4. Connect governor spring and install cover plates.

If the governor contains a torque spring, consult the operator's manual for specific service and adjustment.

TURBOCHARGER

Most compact equipment diesel engines do not contain a turbocharger. However, for those that do, the service required for a turbocharger includes disassembly, cleaning, inspection of parts, and assembly.

Clean the turbocharger and adjacent areas and then remove the turbocharger from the engine by disconnecting mounting bolts, clamps, and oil lines (Fig. 67). Before disassembling the parts, check the turbocharger bearing for radial movement and end play.

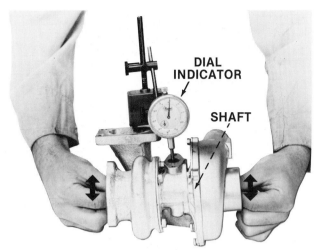

Fig. 68 — Radial Bearing Test

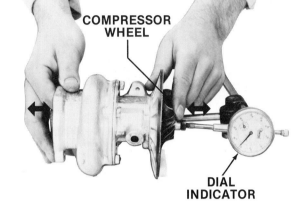

Fig. 70 — Test Bearing End Play

Radial Bearing Test

To make the radial bearing test proceed as follows:

1. Mount a dial indicator on the pump housing with the pin inserted in the lubrication hole for the bearing (Fig. 68). The pin should just touch the turbine shaft.

2. Use both hands to apply side pressure on the shaft. Use equal pressure simultaneously on both ends of the shaft.

3. Check the reading on the dial indicator for the amount of bearing side play.

End Play Bearing Test

Use the following procedure to make the end play bearing test. Partial turbocharger disassembly is required for this test, so place alignment marks at points where the housing will separate (Fig. 69).

1. Remove the compressor housing to expose the compressor wheel.

2. Position a dial indicator with the pin resting on the end of the shaft.

3. Move the shaft back and forth by hand (Fig. 70).

4. Check the reading on the dial indicator for end play.

If radial or end play readings are not within manufacturer's specifications, the bearing must be replaced.

Disassembly

Before disassembly, place alignment marks at points where the turbocharger housing separates to aid in reassembly.

Remove the nuts or screws that hold the turbine and compressor housings to the center housing (Fig. 71). Remove the turbine and compressor housings.

Fig. 69 — Alignment Marks on Turbocharger Housing

Fig. 71 — Screws that Connect Housings

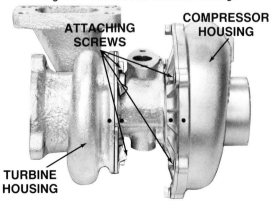

COMPRESSOR WHEEL

RETAINING NUT

TURBINE WHEEL

Fig. 72 — Removing Compressor Wheel Retaining Nut

BLADE EDGES NICKED

Fig. 74 — Compressor Wheel Damaged by Foreign Material

Next, remove the retaining nut (Fig. 72) and remove the compressor wheel. Then withdraw the shaft and turbine wheel assembly (Fig. 73) from the housing.

Disassemble the components following technical manual instructions. Exploded views will also help to keep parts in order. An exploded view of a turbocharger is printed in the Appendix of this text. Other components to disassemble are the seal plate, oil thrower, thrust plate and bushings, and journal bearings.

Cleaning

Before cleaning, inspect all parts for signs of burning, rubbing together, or other damage that may not be detected after parts are cleaned.

Soak all components in clean solvent for about 30 minutes. Then use a stiff bristle brush to remove hardened deposits. Dry all parts with compressed air.

Inspection

Inspect bore surfaces in the center housing for scoring. Use an inside micrometer to measure bearing bores and housing inside diameter. Use a regular micrometer to measure journal bearings and other dimensions specified in the technical manual.

Inspect the turbine shaft for damage from contacting other parts and for heat discoloration. Measure the shaft diameter, the seal ring groove width, and shaft deflection. Replace the shaft if any dimension is not within the limits specified by the manufacturer.

Check for signs of foreign material or dust in the intake or exhaust system by inspecting the turbine and compressor wheels. If damaged by foreign material in the intake system, the compressor wheel may have nicks on the leading edges of the blades (Fig. 74).

Fig. 73 — Removing Turbine Wheel and Shaft

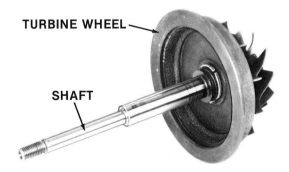

TURBINE WHEEL

SHAFT

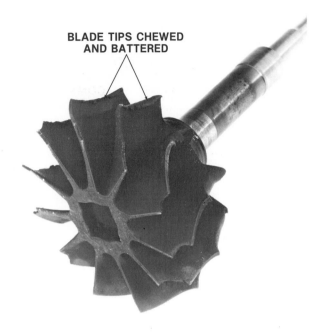

Fig. 75 — Turbine Wheel Damaged by Foreign Material

The turbine wheel blade tips may be chewed and battered (Fig. 75) if damaged by foreign material.

Dust in the intake system will have a sand blasting effect on the compressor wheel. The blades may have nicked leading edges, very thin tips from peening, or eroded blade contours (Fig. 76).

Replace the compressor or turbine wheels if they are worn or nicked. Replace the compressor or turbine housings if they are cracking, rubbing, or warping.

Assembly

Begin turbocharger assembly by installing new inner retaining rings (Fig. 77). Then proceed as follows:

1. Lubricate the journal bearing with engine oil and install into center housing. Install the outer retaining ring to hold the bearing.

Fig. 76 — Sand Blasting Effect on Compressor Wheel

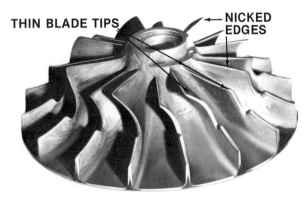

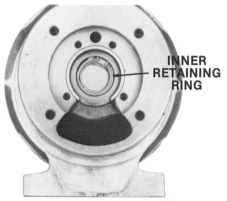

Fig. 77 — Installing Inner Retaining Rings

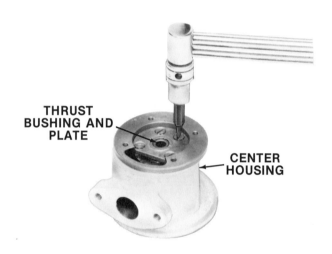

Fig. 78 — Installing Thrust Bushing and Plate

2. Install the thrust bushing and plate onto the center housing (Fig. 78).

3. Install new seal rings onto the oil thrower (Fig. 79) and press the oil thrower onto the seal plate.

4. Coat the surface of the center housing with the recommended sealant and install the seal plate (Fig. 80). Tighten screws to the recommended torque.

5. Install the seal ring on the turbine shaft (Fig. 81).

Fig. 79 — Installing Seal Ring on Oil Thrower

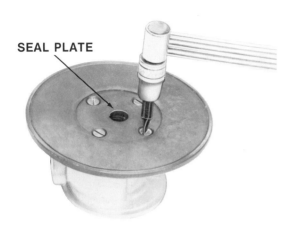

Fig. 80 — Installing Seal Plate

Fig. 82 — Installing Compressor Wheel

6. Install the turbine shaft through the bearing in the center housing and install the compressor wheel on the opposite end of the shaft (Fig. 82).

7. Align the housing locating marks and install the compressor wheel housing (Fig. 83) and turbine wheel housing (Fig. 84). Use the recommended sealant to coat mating surfaces of housings.

Installation

IMPORTANT: Check to see that all parts, tools, and other material are removed from the air intake system before installing the turbocharger.

Use the following procedure to install the turbocharger.

1. Before installing, fill the center housing with clean engine oil. Then rotate the shaft by hand to fully lubricate the journal bearings and thrust washer.

Fig. 83 — Installing Compressor Wheel Housing

Fig. 81 — Installing Seal Ring on Turbine Shaft

Fig. 84 — Installing Turbine Wheel Housing

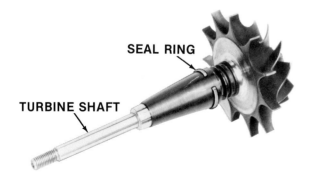

Fig. 85 — Removing Air Cleaner Elements

2. Mount the turbocharger on the engine using a new gasket and tighten mounting bolts.

3. Connect all air hoses and tighten clamps securely. Also connect oil lines and tighten the connections.

4. Start the engine and check the connections for leaks.

COMPONENTS COMMON TO BOTH SYSTEMS

As mentioned earlier in this chapter, several fuel system components are common to both gasoline and diesel fuel systems. These components, which will be covered in the following section, are:

• **Air cleaner**

• **Fuel pump**

• **Fuel tank**

• **Fuel lines**

The components may have a different appearance for different systems, but the service procedures are basically the same.

Fig. 86 — Removing Dust from Element

AIR CLEANER

Two types of air cleaners were discussed in Chapter 4. They were:

• **Dry-element type**

• **Oiled-element type**

The **oiled-element** type air cleaner is more commonly found on compact equipment gasoline engines in the smaller power range. The **dry-element** type air cleaner may be found on any size of engine both gasoline and diesel.

Oil bath air cleaners were not covered in Chapter 4. This type of air cleaner is uncommon in compact equipment. However, oil bath air cleaners are still found on some equipment and the service procedures will be covered.

Dry-Element Type

To service the dry-element type air cleaner, first remove the cover and remove the element (Fig. 85). Both elements shown in Fig. 85 are pleated-paper elements.

If the element is dusty, and the intention is to reuse it, tap the element against the heel of the hand to shake dust out (Fig. 86). Compressed air, with a pressure no greater than 30 psi (200 kPa), can also be used to blow dust from the element (Fig. 86). The compressed air must pass from the inside of the element to the outside.

Fig. 87 — Washing the Element

If the element is oily, and is to be reused, soak and gently agitate the element in a solution of warm water and a commercial filter element cleaner (Fig. 87). Rinse the element thoroughly with clean water, shake out excess water, and allow it to air dry for at least 24 hours.

After cleaning the element, inspect it for damage. Install a new element if the old one is damaged.

Clean the inside of the filter body, and install the element, gasket, and cover.

Oiled-Element Type

Remove the cover and element from the air cleaner body (Fig. 88). Clean the element by washing it in a nonflammable solvent or a solution of detergent and water. Squeeze out all of the cleaning liquid and allow the element to dry completely.

Saturate the element with the same type of oil used to lubricate the engine. Squeeze the element to evenly distribute the oil throughout and then squeeze out excess oil.

Fig. 88 — Assemble Air Cleaner

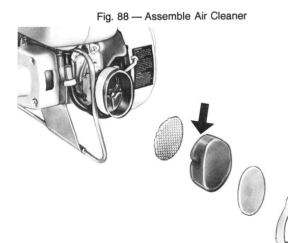

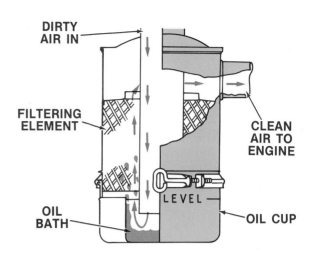

Fig. 89 — Air Flow Through Oil Bath Cleaner

Clean the air cleaner body inside and out and assemble the air cleaner as shown in Fig. 88.

Oil Bath Air Cleaner

Oil bath air cleaners have a cup containing oil attached to the bottom of the air cleaner housing. Air is drawn down a center tube and strikes the surface of the oil in the cup (Fig. 89).

The impact causes the mixture of air and oil spray to be drawn back up through the filtering element. The oil film in the filtering element traps particles of dust and dirt that is in the incoming air.

The particles are then carried back into the cup as the oil drains back. Filtered air continues to flow through the element and on to the engine.

The two main service procedures for oil bath air cleaners are:

- **Maintain the proper oil lever**
- **Change oil**
- **Wash out filtering element**

Periodically check the level of oil in the oil cup. If the level is low, add the recommended oil until the cup is filled to the proper level.

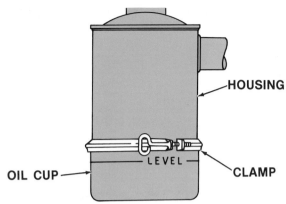

Fig. 90 — Clamp Holds Oil Cup to Housing

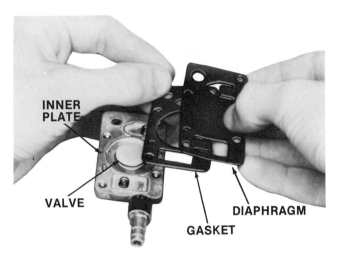

Fig. 91 — Disassembling Fuel Pump

Change the oil in an oil bath air cleaner when the oil appears to have thickened because of excess dirt content. To change the oil, loosen the clamp or bolt which holds the oil cup to the air cleaner housing and remove the cup (Fig. 90).

Pour out the dirty oil and wash out the oil cup with a recommended solvent. Wipe out the cup and the inlet tube with a clean, lint-free cloth.

Fill the oil cup to the OIL LEVEL line with clean engine oil. Install the oil cup back on the housing.

At least once a year, completely disassemble the air cleaner and clean the filtering element. Wash the element in a recommended solvent and allow it to dry completely.

FUEL PUMP

The fuel pump on a gasoline engine and the fuel transfer pump on a diesel engine both perform the same basic function. The same basic service procedures also apply.

Before removing the pump, close the fuel shutoff lever and clean the pump and surrounding area.

Disconnect the fuel inlet and outlet lines from the pump, remove pump mounting bolts, and remove the pump.

Some diaphragm-type fuel pumps are serviceable and some are not. Refer to the technical manual to determine if pump may be repaired.

If the pump is serviceable, whether or not to repair depends on the damage and the cost of repairs. It is often most economical to simply replace the fuel pump if it is not operating properly.

If the pump is to be repaired, disassemble according to instructions in the technical manual (Fig. 91). Remove the diaphragm, gasket, valves, and springs.

Inspect the diaphragm for small holes, wrinkles, and wear. Check the diaphragm mounting surfaces for nicks or burrs. Check for worn valves and broken or distorted springs.

Check for cracking or wear of the gasket, cover, and pump body.

After inspecting the parts, reassemble the pump and install it back on the engine.

FUEL TANK

To service the fuel tank, drain the fuel from the tank and remove the tank from the machine.

Flush the tank for 15 minutes with hot water. Run water into the opening at the bottom of the tank and allow it to overflow.

CAUTION: Inspecting for leaks in the fuel tank is dangerous. Never permit sparks, smoking, or fire of any nature in the area where the tank is being inspected.

Testing

Two methods of testing the tank for leaks are:

- **Wet method**

- **Air pressure method**

WET METHOD

Plug or cap the fuel outlet and fuel return (diesel) tightly. Dry entire outer surface of tank thoroughly with compressed air and a clean, dry rag. Place tank so that all surfaces may be easily seen, such as setting it on top of blocks. Then fill tank with water. Insert end of air hose in filler neck and apply approximately 3 psi (20 kPa) air pressure against water. Examine tank surfaces for moist spots where water may have been forced through.

AIR PRESSURE METHOD

Plug or cap the filler neck and fuel outlet holes. Attach an air hose to fuel outlet. Submerge fuel tank in clean water and apply approximately 3 psi (20 kPa) air pressure. Draw a ring around each spot on fuel tank where bubbles appear. These bubbles indicate leaks in tank that need repairing.

Repair

Many fuel tanks on compact equipment are not repairable. If this is the case, replace it with a new one.

FUEL LINES

Low pressure fuel lines may be made of steel tubing or a polyethylene material. High pressure lines in the diesel fuel system, are heavier than others but service procedures are similar for all lines.

The only inspecton and service required for fuel lines is to check the lines and connections for fuel leakage. If leakage is detected, replace that portion of the line.

Also check for and replace bent or dented fuel lines which could restrict fuel flow.

SUMMARY

This chapter discusses service procedures for the following:

- **Gasoline fuel system components**
- **Diesel fuel system components**
- **Components common to both systems**

Service procedures for fuel system components consist of:

- **Disassembly**
- **Cleaning**
- **Inspection and repair**
- **Reassembly and installation**
- **Adjustment (if required)**

GASOLINE FUEL SYSTEM COMPONENTS

The different types of **carburetors** all require some different service procedures. However, many of the procedures are similar such as disconnecting fuel lines and linkages, removing needle valves, and removing choke and throttle valves.

After parts are disassembled, clean and inspect them. Replace parts with those offered in a carburetor kit. Reassemble the carburetor and make adjustments to needle valves.

Governor service procedures also vary depending on the design of the governor. Basic service consists of disassembling parts and inspecting them for wear.

The parts to be disassembled depend on whether the **governor** is an **air-vane or flyweight** type. After parts are inspected and damaged parts are replaced, reassemble the governor and install it on the engine. Make sure that linkages are not binding.

DIESEL FUEL SYSTEM COMPONENTS

The **fuel injection pump** may be removed and replaced but should not be disassembled in the service department unless the department has qualified technicians and the necessary tools and test stand. All fuel lines, linkages, and mounting bolts must be removed for disassembly.

Injection pump adjustment consists of timing the pump so that injection occurs at the proper time.

When the diesel fuel system has been opened for service, air can enter the system. This air must be bled from the system for the engine to run efficiently. Bleed air at the filter, pump, or injection nozzle.

The **fuel injection nozzle** may be disassembled, cleaned, and inspected.

Diesel fuel filters may contain a built-in or a separate water separator. Diesel fuel filtration is especially important because of the precision-made parts of the diesel engine. The parts are more susceptible to damage from contamination and water than gasoline engine parts.

The **governor** in some diesel fuel systems is not serviceable by all service departments. The governor should only be serviced by a qualified technician who has the necessary tools and equipment.

If the governor is serviceable, remove mounting bolts and disassemble the components. Governors that are serviceable may also be adjusted. Follow the procedure recommended in the technical manual.

Turbochargers are not common on compact equipment diesel engines. However, some equipment models use a turbocharger.

Test the turbocharger bearings for radial and end play. Then disassemble, clean, and inspect turbocharger parts.

COMPONENTS COMMON TO BOTH SYSTEMS

Components that are common to both gasoline and diesel fuel systems and required similar servicing are:

- **Air cleaner**
- **Fuel pump**
- **Fuel tank**
- **Fuel lines**

CHAPTER 7 REVIEW

1. (Fill in the blanks) The most common types of carburetors on compact equipment engines are _____ type, _____ - _____ type, and _____ type.

2. (Multiple choice) At what time would a carburetor **not** likely require adjustment:

    A. After a major engine overhaul
    B. Each day when the engine is started
    C. During engine tune-up
    D. Whenever the carburetor has been removed for service

3. (True or false) If some carburetor passages will not come clean by soaking in solvent, use a small drill bit or a wire to clean the passage.

4. (Fill in the blanks) Carburetors with adjustable mixture needles must be adjusted for _____ _____ and _____ operation.

5. (True or false) When the engine is stopped, the governor spring should hold the throttle valve closed.

6. Describe the basic steps required to remove a fuel injection pump.

7. (Fill in the blank) If air enters the diesel fuel system, _____ _____ can form which could prevent fuel from reaching the pump or nozzle.

8. (Fill in the blanks) On a three-cylinder diesel engine, adjust injection pump timing on the No. _____ cylinder.

9. (Multiple choice) The three tests normally performed on fuel injection nozzles are (select three):

    A. Compression test
    B. Leakage test
    C. Opening pressure test
    D. Chatter and spray pattern test
    E. Air-in-nozzle test

10. (Fill in the blanks) The procedure used to check the amount of side play of the turbocharger bearing is called the _____ _____ test.

11. Describe the procedure to use to clean a dry-element type air cleaner.

12. (Fill in the blanks) The two methods used to check the fuel tank for leaks are _____ and _____ _____ methods.

CHAPTER 8

SERVICING THE COOLING SYSTEM

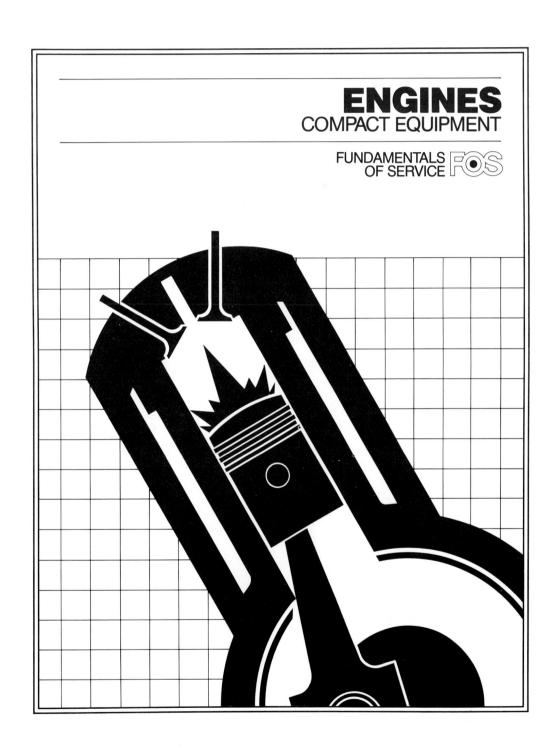

ENGINES
COMPACT EQUIPMENT

FUNDAMENTALS
OF SERVICE FOS

SKILLS AND KNOWLEDGE

This chapter contains basic information that will help you gain the necessary subject knowledge required of a service technician. With application of this knowledge and hands-on practice, you should learn the following:

• Service procedures for cooling fins on air-cooled engines

• Procedures for inspecting and leak testing the radiator and radiator cap

• How to inspect the fan and fan belt and how to check the condition of the fan belt

• Removal, disassembly, and assembly of the water pump

• Inspection procedures for radiator hoses

• Procedures for servicing the entire cooling system

• How to inspect the cooling system for leaks

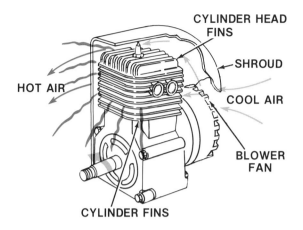

Fig. 1 — Cooling Fins Radiate Heat

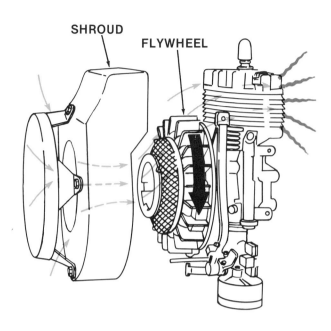

Fig. 2 — Engine Shroud Directs Cooling Air Flow

INTRODUCTION

As mentioned in Chapter 1 of this text, compact equipment engines may have one of two types of cooling systems. The two types are:

- **Air cooling**

- **Liquid cooling**

The components and operation of both systems were covered in Chapter 1. This chapter will briefly summarize the components of each system and then describe service procedures to keep the cooling system operating efficiently.

AIR COOLING

An air-cooled engine is equipped with cooling fins on the cylinder head, cylinder, and cylinder block (Fig. 1). These fins radiate heat from the engine into the surrounding air.

Air-cooled engines often contain a fan or blower to increase air flow to aid in the heat transfer. This fan is usually part of the flywheel, but it may be a unit mounted separately from the flywheel. In both cases, shrouds on the engine are used to direct air flow where it is most needed (Fig. 2).

SERVICING THE AIR-COOLED ENGINE

Air-cooled engines have a simple method of cooling — hot air is transferred from the engine to the surrounding air. Because of their simplicity, service procedures for air-cooled engines are relatively simple.

Before disassembling, clean any dirt, trash, or oil from the exterior of the engine. Then remove the shrouds that are used to protect the engine and direct air flow from the fan (Fig. 3).

After the shrouds are removed and the cylinder head, cylinder, and cylinder block are exposed, inspect the cooling fins for any buildup of trash or dirt between the

Fig. 3 — Remove Engine Shroud

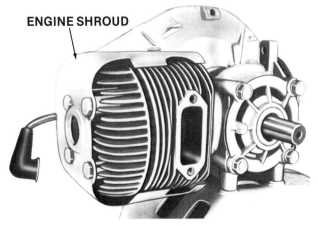

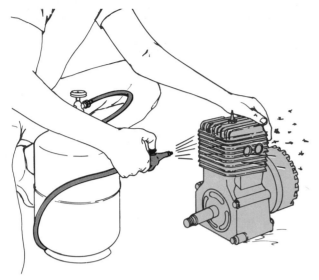

Fig. 4 — Clean Exterior of Engine

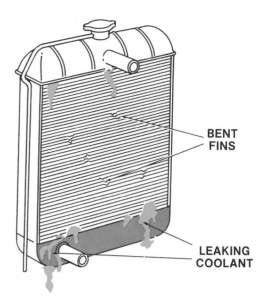

Fig. 6 — Bent Fins on Radiator

fins. Any material of this sort can trap heat in the engine and cause the engine to overheat.

If material must be cleaned from the cooling fins, allow the engine to cool to ambient temperature. Then use compressed air or a pressure washer to clean the exterior of the engine (Fig. 4).

After the engine is cleaned, reinstall the shrouds.

Check for broken fins on the fan or the fan portion of the flywheel. Broken fins could cause vibrations and also could result in engine overheating.

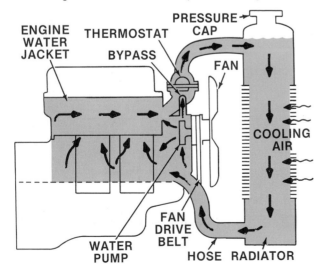

Fig. 5 — Water Jacket on Liquid-Cooled Engine

LIQUID COOLING

A liquid-cooled engine is equipped with a water jacket and passages around the cylinder and through the cylinder head (Fig. 5). The parts of a typical liquid-cooling system are:

- **Radiator**

- **Fan and fan belts**

- **Water pump**

- **Radiator hoses**

- **Thermostat**

The following discussion will cover the service procedures for each of these parts. If the engine fails to operate properly, and the cooling system is suspected as the cause, read Chapter 10 of this text for engine and cooling system diagnostics. Various cooling system tests are also covered in Chapter 10. These tests include:

- **Leak test entire system**

- **Radiator cap test**

- **Thermostat test**

- **Exhaust gas leakage test**

- **Air-in-system test**

RADIATOR

Inspect the radiator for bent fins (Fig. 6) and straighten those that are bent. Inspect the inlet and outlet tubes for cracks, kinks, dents, and fractured seams.

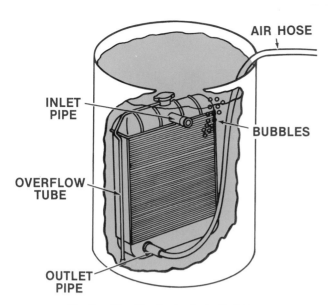

Fig. 7 — Checking for Leaks in Radiator

Check for dirt and insects that may lodge in the radiator. Clean them out by using compressed air or a pressure washer.

If a radiator leak is detected, but cannot be spotted visually, test it as follows:

1. Install the radiator cap and plug the overflow tube and outlet pipe. Attach an air hose to the inlet connection.

2. Pressurize the inside of the radiator with compressed air to about 7 to 10 psi (50 to 70 kPa).

3. Submerge the radiator in water and look for bubbles which indicate the location of the leak (Fig. 7).

4. Repair any leaks by soldering.

Radiator Cap

The radiator cap has two functions:

• *Allows atmospheric pressure to enter the cooling system*

• *Prevents coolant escape at normal operating pressure*

A **pressure valve** in the cap permits the escape of coolant or steam when the temperature rises above a certain point (Fig. 8).

A **vacuum valve** in the cap opens when needed to prevent a vacuum in the cooling system (Fig. 8).

The radiator cap must be removed from the radiator to check the valves.

CAUTION: Always allow the engine to cool before removing the radiator cap. Then remove the cap slowly and carefully to avoid a possible fast discharge of hot coolant which could cause severe burns.

To prevent cooling system damage from either excessive pressure or vacuum, check both valves periodically for proper opening and closing pressures. See Chapter 10 for procedures for testing the radiator cap.

Fig. 8 — Valves in Radiator Cap

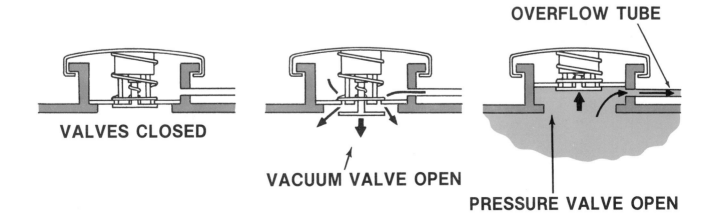

VALVES CLOSED

VACUUM VALVE OPEN

OVERFLOW TUBE

PRESSURE VALVE OPEN

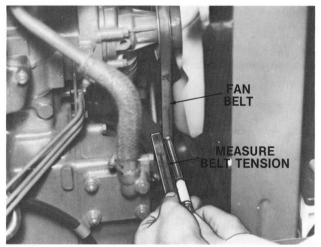

Fig. 9 — Checking Belt Tension

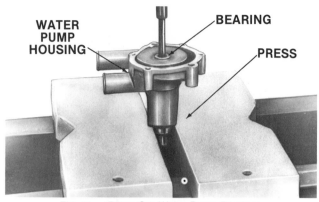

Fig. 11 — Press Out Water Pump Bearings

FAN AND FAN BELTS

The only service on the engine **fan** is to be sure the blades are straight and far enough from the radiator to not strike the core. Bent blades reduce the cooling system efficiency and throw the fan out of balance.

The **fan belt** should be neither too tight nor too loose. A tight belt puts an extra load on the fan bearings and shortens the life of the bearings and the belt. A loose belt allows belt slippage and lowers the fan speed. This causes excessive belt wear and leads to cooling system overheating.

Check the belt condition and belt tension (Fig. 9) periodically. Adjust the tension if the belt is too tight or too loose.

WATER PUMP

When the water pump fails to circulate the coolant, the engine will likely overheat and be damaged. If the water pump fails to circulate the coolant properly, it must be removed and repaired or replaced.

Removal And Disassembly

To remove the water pump:

1. Remove coolant hoses from the pump (Fig. 10)

2. Remove any shrouds

3. Remove the fan belt

4. Remove the capscrews or bolts that hold the water pump on the engine and remove the water pump.

Disassemble the pump by first removing any retainer rings that hold the pump parts in place.

Fig. 10 — Coolant Hose Attached to Water Pump

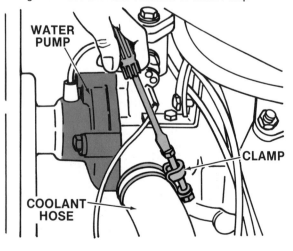

Fig. 12 — Water Pump Parts

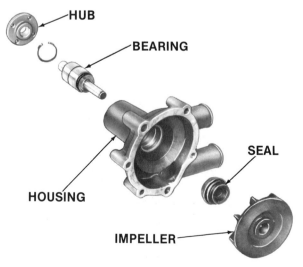

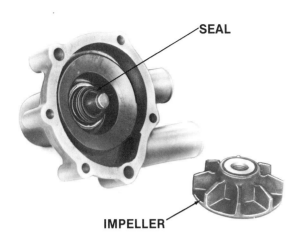

Fig. 13 — Water Pump Impeller and Seal

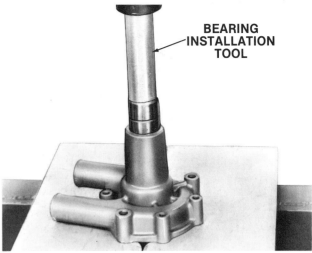

Fig. 15 — Installing Bearing

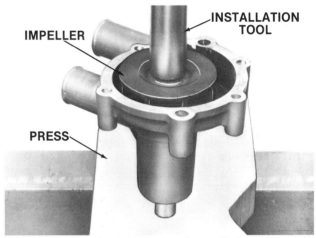

Fig. 16 — Installing Impeller on Shaft

Then support the water pump in a press (Fig. 11). Press the bearings from the impeller and water pump housing. Remove other parts from the housing (Fig. 12).

Inspection And Assembly

Inspect the parts for wear, particularly the seal and insert in the impeller (Fig. 13).

To reassemble the water pump:

1. Install the seal by using a hydraulic press and a socket the same diameter of the seal (Fig. 14).

2. Use a special tool to press in the water pump bearing until it is properly seated (Fig. 15).

3. Support the pump housing in a press and press the impeller onto the water pump shaft (Fig. 16).

4. Press the hub onto the water pump shaft (Fig. 17).

Fig. 17 — Press Hub onto Shaft

Fig. 14 — Installing Seal

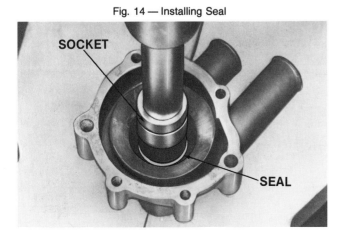

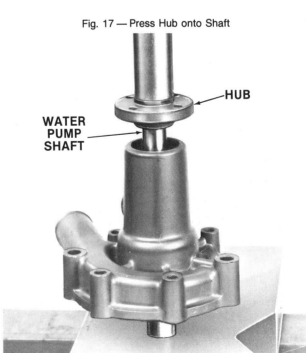

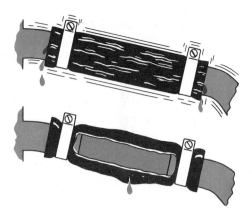

Fig. 18 — Leakage Points on Hoses

After the water pump is assembled, install the pump back on the engine. Put all belts, hoses, and shrouds back into position.

RADIATOR HOSES

Flexible hoses are used to connect many cooling system components because they can withstand vibration better than rigid pipes. Hoses, however, can be damaged by air, heat, and water in two ways:

• **Hardening or cracking** — Destroys flexibility, causes leakage, and allows small particles of rubber to clog the radiator.

• **Softening and swelling** — Produces lining failure and hose rupture.

Check hose condition at least twice a year. Make sure they are pliable and able to pass coolant without

Fig. 19 — Hose Interior Worn Out

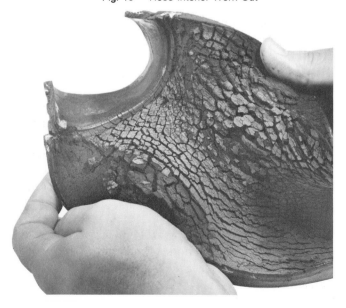

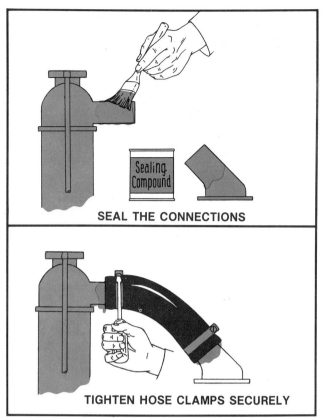

SEAL THE CONNECTIONS

TIGHTEN HOSE CLAMPS SECURELY

Fig. 20 — Installing Hoses

leakage (Fig. 18). Also check any reinforcing springs inside the hoses for corrosion.

Hoses can deteriorate on the inside and still appear to be in good condition on the outside (Fig. 19). When hoses are removed, check them for wear. Hoses may harden or crack allowing the system to leak, or they may soften and then collapse. A softened hose may also collapse during high-speed operation. A collapsed hose can restrict coolant circulation enough to cause engine overheating.

Installing Hoses

To install hoses, first clean the pipe connections and apply a thin layer of a sealing compound to the connections (Fig. 20).

Locate the hose clamps properly over the connections and tighten the clamps. A pressurized cooling system will blow a hose that is not installed properly.

THERMOSTAT

Before removing the thermostat for service, check the area around the thermostat for coolant leaks. If leaks

Fig. 21 — Thermostat Cover

are detected, inspect the thermostat cover for cracks or check for a leaking gasket after the cover is removed.

Next, remove the radiator hose from the thermostat cover and then remove the cover (Fig. 21). Remove the thermostat from its housing (Fig. 22).

Inspecting Thermostats

If a thermostat is broken or corroded, discard it.

If it is not of the proper temperature and type, as indicated on an application chart, replace it. The number stamped on it is the approximate temperature that it will reach before starting to open.

Test the thermostat according to procedures described in Chapter 10:

Installing Thermostats

When installing a thermostat in the engine water jacket, position the thermostat with the expansion element toward the engine.

Fig. 22 — Remove Thermostat

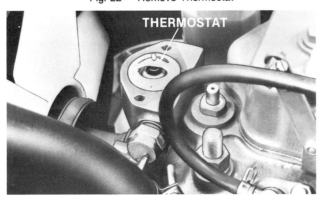

Some thermostats are marked with arrows that point to the radiator or to the engine block, or are marked "top" or "T". "Front" is indicated on some models. **The frame must not block the water flow.**

To prevent leakage, clean the gasket surfaces on the thermostat. Use a new gasket; normally it does not require a sealing compound.

When the cover and gasket have been properly located, tighten the nuts evenly and securely.

SERVICING THE LIQUID COOLING SYSTEM

Correct cooling system servicing is vital for a smooth-running engine.

Overheating is a big problem and it can be caused by:

- **Lack of coolant**
- **Loose or slipping belts**
- **Clogging of cooling system**
- **Malfunctioning water pump or thermostat**

Check the coolant level and temperature frequently. Service the entire cooling system at least twice a year.

Cleaning The Cooling System

Efficient operation of the cooling system requires that it be cleaned occasionally, particularly at seasonal changes when antifreeze solution is added or removed.

Several methods of cleaning the system are available, the proper one depending upon the amount of corrosion present in the system.

Rust-Proofing The System

Rust is formed by corrosion of iron parts:

IRON + WATER + OXYGEN = RUST

Check the cooling liquid spring and fall. For best system performance, drain the system at least once a year.

Use a cooling system cleaner if the liquid drained is rusty. Otherwise, use a radiator flush or plain water. Corrosion inhibitors do not clean out rust already formed.

In the spring, if the antifreeze is drained, renew the rustproofing by adding a recommended antirust compound to the fresh water, or by installing new antifreeze.

Regardless of the summertime practice, corrosion protection can best be provided in winter by inhibitors built into the antifreeze coolant.

IMPORTANT: Adding rust inhibitors or fresh antifreeze to used solutions will not restore full strength corrosion protection. Some mixtures may even do harm.

Some manufacturers recommend the year-round use of a highly inhibited antifreeze coolant. However, remember that the service life of the inhibitor may be drastically reduced as the machine gets older and the coolant may have to be replaced more often.

Some of the specific factors involved are:

• *Accumulation of contaminants, including rust and corrosion products.*

• *Dilution with water that was added to replace leakage losses, and possible absorption of air and exhaust gases that contain water.*

For these reasons, be sure to:

1. Check the cooling system and coolant periodically.

2. Maintain at least a 25 percent solution concentration year-round for adequate corrosion protection.

3. Install new antifreeze at least once each year.

Clogging

Clogging is a fairly common cause of cooling system trouble. It can be avoided entirely by periodic rustproofing and cleaning when necessary.

The most common clogging materials are:

• **Rust**

• **Scale**

• **Grease**

• **Lime**

Rust accounts for 90 percent of the clogging. It forms on the walls of the engine water jacket and other metal parts.

Grease and oil enter the system through:

• **Cylinder head joint**

• **Water pump**

Coolant circulation loosens rust particles. The particles settle in the water jacket and build up in layers inside the radiator water tubes.

As the rust layer becomes thicker it keeps cutting down heat transfer from the radiator until the engine overheats and boiling starts in the water jacket.

The boiling stirs up more rust in the block and forces it into the radiator — eventually clogging it.

Mineral Deposits

Mineral deposits form rapidly at hot spots in the engine (Fig. 23).

Overheating, knock, and eventually engine damage can result from the buildup of rust and mineral scale on the water side of the combustion chamber (Fig. 23).

To avoid excessive formation of rust or scale deposits:

• *Keep the cooling system free of leaks*

• *Avoid adding too much hard water*

• *Maintain full strength corrosion protection at all times*

Cooling System Cleaners

Well maintained cooling systems seldom, if ever, require corrective cleaning.

However, if periodic rustproofing or other preventive maintenance is neglected, deposits will build up and cleaners must be used to restore cooling capacity.

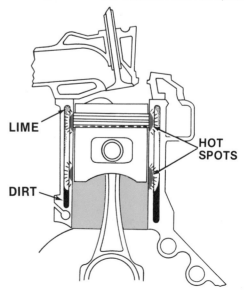

Fig. 23 — Mineral Deposits Form Hot Spots

Use a cleaner to remove:

- **Hard rust deposits**

- **Scale**

- **Grease**

Rust and water scale build up in the cooling system. Clogging material is usually made up of grease, rust, and water scale, but is 90 percent iron rust.

Remove both hard rust scale and grease with a recommended double-acting cleaner. It is harmless to cooling system metals and hose connections when used according to directions.

IMPORTANT: If rust and grease are not completely neutralized and flushed out, they can destroy the corrosion inhibitors in later fills of antifreeze or antirust solutions.

NOTE: Some radiators are made of aluminum or other metals which may be affected by certain compounds used for immersion cleaning. Follow manufacturers' instructions for cleaning these radiators.

Flushing The Cooling System

Incomplete flushing, such as hosing out the radiator, closes the thermostat and prevents thorough flushing of the water jacket. For complete flushing, take the following steps.

1. Fill the system completely with fresh water.

2. Run the engine long enough to open the thermostat (or remove the thermostat).

3. Open all drain points to drain the system completely.

Clean out the *overflow pipe* and remove insects and dirt from radiator air passages, radiator grill and screens.

Also check the thermostat, radiator pressure cap, and the cap seat for dirt or corrosion.

Cooling System Leakage

Leakage is the most common problem in a cooling system. During the winter it can result in the loss of valuable antifreeze.

But leakage actually does increase during winter due to metal shrinkage and cooling system pressure.

Air pressure leakage testers can help locate external leaks but they cannot be depended upon to locate small combustion leaks.

Only practical experience enables a service technician to tell whether the location and size of a leak can be corrected with a sealing solution.

All leakage exposed to combustion pressures must be repaired mechanically.

Follow instructions when using sealing solutions. Some sealing solutions react chemically with antifreeze and rust inhibitors, and seriously affect coolant performance.

RADIATOR LEAKAGE

Most radiator leakage is due to mechanical failure of soldered joints. This is caused by:

- **Engine vibration**

- **Frame vibration**

- **Pressure in the cooling system**

Examine radiators carefully for leaks before and after cleaning. Cleaning may uncover leakage points already existing but plugged with rust.

White, rusty or colored leakage stains indicate previous radiator leakage. These spots may not be damp if water or alcohol is used since such coolants evaporate rapidly, but ethylene glycol antifreeze shows up because it does not evaporate.

Always stop any radiator leakage before installing antifreeze coolant.

LEAKAGE OUTSIDE ENGINE WATER JACKET

Inspect the engine cylinder block before and after it gets hot while the engine is running.

Leakage of the engine block is aggravated by:

- **Pump pressure**

- **Pressurized cooling systems**

- **Temperature changes of the metal**

Remember: Small leaks may appear only as rust, corrosion or stains, due to evaporation.

Watch for these trouble spots:

1. **Core-Hole Plugs.** Remove old plug. Clean plug seat, coat with sealing compound. Drive new plug into place with proper tool.

2. **Gaskets.** Tighten joint or install new gasket. Use sealing compound when required.

3. **Stud Bolts And Cap Screws.** Apply sealing compound to threads.

WATER JACKET LEAKAGE INTO ENGINE

Coolant leaks into the engine through:

* *A loose cylinder head or sleeve joint*
* *A cracked or porous casting*
* *The push rod compartment*

Water or antifreeze mixed with engine oil will form **sludge** which causes:

* *Lubrication failure*
* *Sticking piston rings and pins*
* *Sticking valves and valve lifters*
* *Extensive engine damage*

The amount of damage depends upon the amount and duration of leakage, service procedures, and seasonal operating conditions.

High temperature thermostats allow the engine heat to remove moisture (formed from blowby) from the crankcase.

Head gasket leakage is more likely to occur in winter than summer because of increased metal contraction and expansion in winter.

Give special attention to the cylinder head gasket no matter what type or material.

An improperly installed gasket can cause:

* **Coolant and oil leakage**
* **Overheating**

It may be necessary to pressurize the cooling system and tear down part of the upper part of the engine to find a coolant leak in the push rod compartment.

Replacing Head Gaskets

To replace a cylinder head gasket:

1. Use only a new gasket designed for the engine.

2. Make sure the cylinder head and block surfaces are clean, level and smooth.

3. See Chapter 5 for details.

IMPORTANT: Leading engine manufacturers strongly recommend the use of torque wrenches to tighten cylinder head bolts.

Always follow the engine manufacturers' instructions on cylinder head bolt tension and order of tightening.

Overheating Damage

The intense heat of combustion can cause engine components such as valves, pistons and rings to operate near their critical temperature limits — even when the cooling system is operating normally.

Overheating severely affects the lubrication of the engine. High metal temperatures can destroy the lubricating film, accelerate oil breakdown, and cause formation of varnish.

These in turn may cause:

* **Excessive wear**
* **Scoring**
* **Valve burning**
* **Seizure of moving parts**

Continued operation above the normal temperature range may result in:

* **Lubrication failure**
* **Heat distortion of parts**
* **Engine knocking**

Cylinder heads and engine blocks are often warped and cracked (Fig. 24) by strains set up in the metal by overheating, especially when followed by rapid cooling.

Hot spots in an overheated engine can cause engine knocking. If allowed to continue, knocking results in:

* **Blown head gaskets**
* **Damaged pistons**
* **Ring and bearing failure**

Exhaust Gas Leakage

A cracked cylinder head or a loose cylinder head joint allows hot exhaust gas to be blown into the cooling

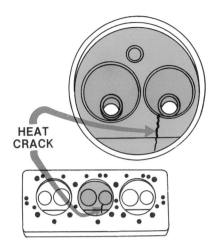

Fig. 24 — Heat Crack in Cylinder Head

system under combustion pressures; even though the joint may be tight enough to keep liquid from leaking into the cylinder.

The cylinder head gasket itself may be burned and corroded by escaping exhaust gases.

Exhaust gases dissolved in coolant destroy the inhibitors and form acids which cause corrosion, rust and clogging.

Excess pressure may also force coolant out the overflow pipe.

Corrosion In The Cooling System

Water solutions will corrode the cooling system metals when they are not protected by inhibitors added to water for summer operation.

Corrosion coating on metal surfaces reduces heat transfer at the walls of the engine water jacket and in the tubes of the radiator even before clogging occurs.

Galvanic action is a form of corrosion which can occur when two different metals (in contact) are suspended in a liquid which will carry a current.

It is similar to what happens in a battery when electric current flows from one area to another, removing metal from one electrode and depositing it on another in the process.

Galvanic corrosion or direct attack by the coolant will affect the following metals in the cooling system: stainless steel, brass, solder, aluminum, and copper.

Aeration In Cooling System

Aeration, or mixing of air with water, speeds up the formation of rust and increases corrosion of cooling system metals.

Aeration may also cause:

- **Foaming**
- **Overheating**
- **Overflow loss of coolant**

Air may be drawn into the coolant because of:

- **A leak in the system**
- **Turbulence in the top tank**
- **Too low coolant level**

Check the cooling system for exhaust gas leakage and air suction when you find these conditions:

- **Rusty coolant**
- **Severe rust clogging**
- **Corrosion**
- **Overflow losses**

CHAPTER 8 REVIEW

1. Discuss the service procedures required for an air-cooled engine.

2. (Fill in the blanks) The radiator cap contains two valves known as the _____ valve and the _____ valve.

3. Why is it necessary that the blades of the engine fan are kept straight?

4. (True or false) Radiator hose hardening and cracking will likely result in lining failure and hose rupture.

5. (True or false) The addition of corrosion inhibitors will clean out rust that has formed in a cooling system.

6. (Multiple choice) Which of the following is **not** recognized as one of the materials that most frequently clog the cooling system:

   A. Grease
   B. Rust
   C. Dust
   D. Scale
   E. Lime

7. (Fill in the blank) _____ is the most common problem in a cooling system.

8. What are three main causes of radiator leakage?

9. The presence of air or exhaust gas in the coolant may show up as one of four different conditions in the system. What are the four conditions?

CHAPTER 9

SERVICING THE LUBRICATION SYSTEM

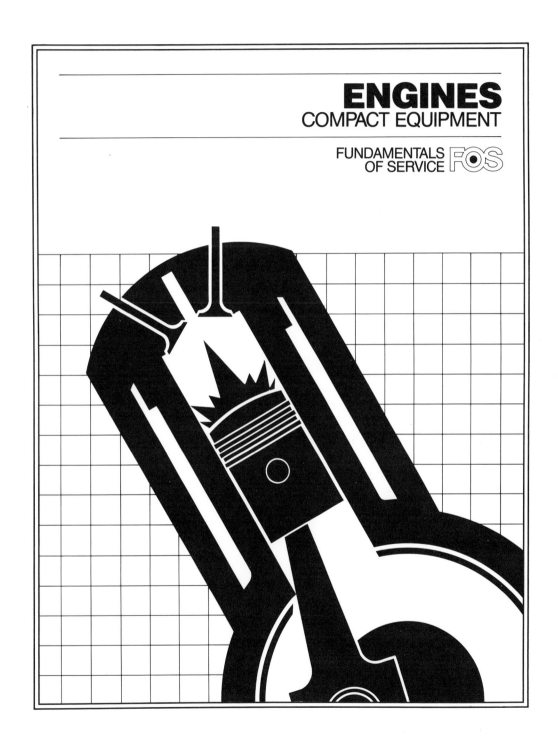

ENGINES
COMPACT EQUIPMENT

FUNDAMENTALS
OF SERVICE FOS

SKILLS AND KNOWLEDGE

This chapter contains basic information that will help you gain the necessary subject knowledge required of a service technician. With application of this knowledge and hands-on practice, you should learn the following:

• How to remove and service the plunger, rotor, and external-gear oil pumps

• How to remove, inspect, and adjust the oil pressure regulating valve

• Tips on keeping contaminants out of oil

• How to protect the engine against oil contamination

• Causes of oil consumption

• How to change engine oil and oil filter

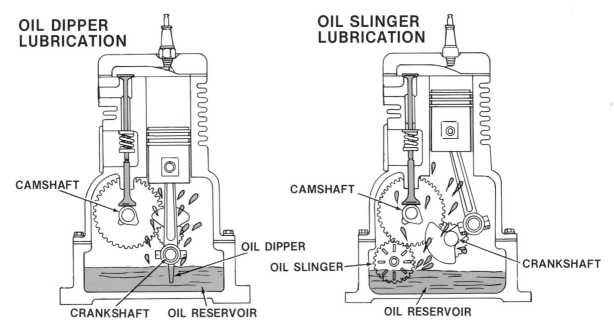

Fig. 1 — Splash-Type Lubrication System

INTRODUCTION

Most two-cycle engines contain no lubrication system components because the lubricating oil is mixed with the gasoline. Oil injected 2-cycle engines do contain some lubrication system components. However, engines with oil injection are generally high performance engines and are beyond the scope of this text. Therefore, 2-cycle engine lubrication systems are not covered in this chapter.

Lubrication systems on 4-cycle engines require careful servicing to prevent failure of engine parts. Engine parts could fail due to a lack of lubrication or from lubrication with dirty oil.

The two basic types of 4-cycle engine lubrication systems are:

- **Splash type (Fig. 1)**

- **Pressure type (Fig. 2)**

In both systems, maintaining the proper oil level is probably the most important service procedure to perform.

The **splash-type system** is relatively simple and requires little service other than checking and changing the oil. Inspect the engine parts during any disassembly to make sure all parts are in good working condition.

The **pressure-type system**, however, is more complex, contains more components, and, therefore, requires more servicing. The remainder of this chapter will cover service procedures for pressure-type lubrication systems on 4-cycle engines.

The components of the 4-cycle engine pressure lubrication system which require service are (Fig. 2):

- **Oil pump**

- **Oil filter**

- **Pressure regulating valve**

Service procedures for these components will be discussed in the following sections.

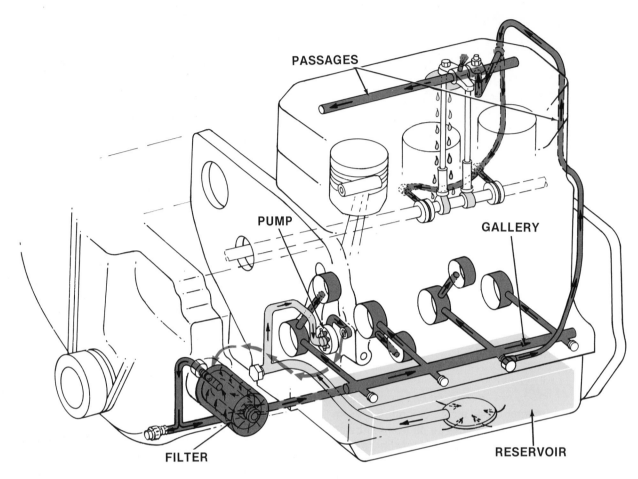

Fig. 2 — Pressure-Type Lubrication System

OIL PUMP

The three types of oil pumps most commonly found in pressure lubrication systems are:

- **Plunger pump**
- **Rotor pump**
- **External-gear pump**

The service and repair procedures vary for each type.

PLUNGER PUMP

The plunger-type oil pump is located inside the engine (Fig. 3). Remove the end wall of the crankcase, as explained in Chapter 5, to gain access to the pump.

Remove the pump by lifting it from the shaft on which it is mounted (Fig. 3). Then remove the plunger from the barrel. Clean the parts in a recommended solvent.

Inspect the parts for wear or damage and replace the pump if it is worn.

Check the pump for proper operation by submerging it in oil and operating it. Make sure oil is being pumped from the outlet.

Fig. 3 — Plunger-Type Oil Pump

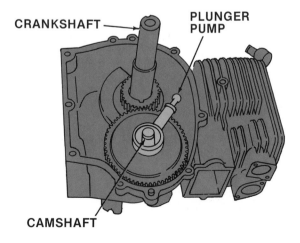

Fig. 4 — Checking Oil Pump Gear Backlash

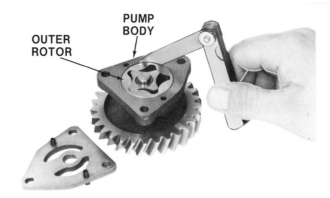

Fig. 6 — Checking Clearance between Rotor and Pump Body

After checking the pump, install it back on the shaft and reassemble the engine.

ROTOR PUMP

The rotor pump is often driven by a gear that is in mesh with the crankshaft gear or an idler gear. Before removing the oil pump for service, check the oil pump gear backlash.

To do this, first remove the timing gear cover to expose the oil pump gear. Install a dial indicator on the engine block as shown in Fig. 4. Move the oil pump gear back and forth and observe the reading on the dial indicator. While moving the gear, make sure that it is not moved so far as to cause the crankshaft gear to move. This would give a false reading on the dial indicator.

Compare the reading on the dial indicator to the manufacturer's specification. If the reading exceeds the wear tolerance, replace the oil pump gear.

Remove the capscrews that hold the pump to the engine and remove the pump.

Inspection

Check the recess of the pump rotors by placing a straightedge across the machined surface of the pump body and measuring with a feeler gauge (Fig. 5). Replace the pump if the wear is not within the specified tolerance.

Use a feeler gauge to check the clearance between the outer rotor and the pump body (Fig. 6). Again, if the wear is more than the specified wear tolerance, replace the pump.

Check the clearance between the inner rotor and outer rotor with a feeler gauge (Fig. 7). Measure the clearance between a high point of the inner rotor and a high point of the outer rotor. Again, replace the pump if the measurement exceeds the wear tolerance.

Assembly And Installation

Assemble all parts of the rotor pump (Fig. 8) in the following sequence.

Fig. 5 — Checking Recess of Pump Rotors

Fig. 7 — Checking Clearance between Pump Rotors

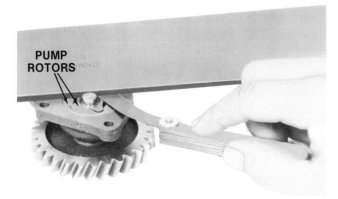

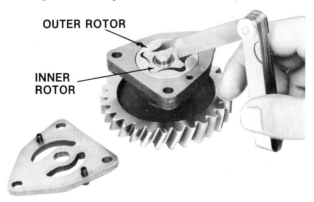

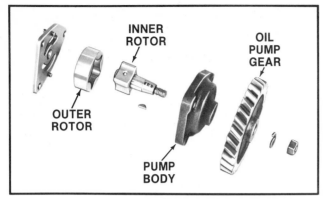

Fig. 8 — Parts of the Rotor Pump

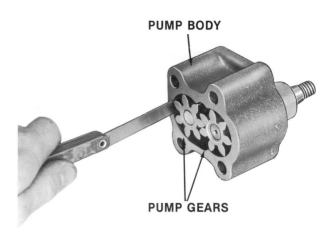

Fig. 10 — Measuring Clearance between Gears and Housing

1. Install the inner rotor-shaft assembly in the pump body.

2. Install the outer rotor in the pump body.

3. Install the key in the inner rotor-shaft assembly.

4. Place the capscrews in the pump body.

5. Install the gear on the shaft.

6. Install the lock washer and nut on the shaft and tighten.

7. Install the pump on the engine (Fig. 9).

EXTERNAL-GEAR PUMP

To service the external-gear pump, first remove the capscrews that hold the pump to the engine. Remove the pump from the engine.

Most wear in an external-gear pump will occur between the teeth of the gears.

Inspection

Use a feeler gauge to measure the clearance between the gears and the housing (Fig. 10). If clearance is not within the specified wear tolerance, replace the gears.

Place a straightedge across the top of the pump housing (to represent the cover) and measure the clearance between the gears and the straightedge (Fig. 11). If the clearance is excessive, replace the gears.

Almost all pumps use a screen over the inlet to strain out foreign material. Where possible, remove the screen from the inlet pipe, and clean both with a solvent. Use compressed air to dry the parts.

Inspect all bushings in the housings and replace those with excessive wear.

OIL FILTER

Oil filters are used to filter contaminants from the oil before the oil is allowed to reach engine parts.

Fig. 9 — Installing Pump on Engine

PUMP MOUNTING BOLTS

Fig. 11 — Measuring Gear Side Clearance

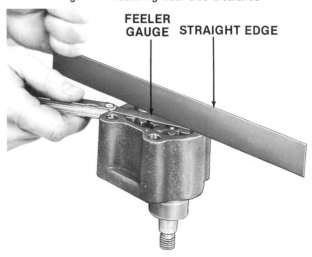

FEELER GAUGE   STRAIGHT EDGE

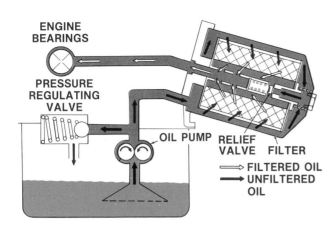

Fig. 12 — Oil Filter with Bypass Valve

Fig. 14 — Pressure Regulating Valve Body

The oil filter cannot be cleaned. Change the filter at intervals recommended in the technical manual. Change engine oil at the same time the filter is changed.

Some oil filters contain a filter bypass valve (Fig. 12). The bypass valve operates during cold weather starting when the viscosity of the oil may be too thick to pass easily through the filter. It also operates when the oil filter becomes plugged and will not allow oil to pass through.

When the bypass valve opens, oil simply bypasses the filter and lubricates engine parts.

PRESSURE REGULATING VALVE

Before checking engine oil pressure, check the condition of the oil filter. A dirty filter can limit the flow of oil.

Check the operation of the pressure regulating valve by checking engine oil pressure. This test is covered in Chapter 10.

Fig. 13 — Pressure Regulating Valve Under Oil Filter

The pressure regulating valve on some engines is located under the oil filter. On these engines, first remove the oil filter (Fig. 13).

Then remove the pressure regulating valve body and the spring retainer from the valve body (Fig. 14). Also, remove and inspect the shim(s), check ball, and spring (Fig. 15).

Adjust oil pressure by adjusting the number of shims in the valve body. Generally, to **raise oil pressure**, add shims behind the spring to increase spring tension. This increases the setting at which the valve will open.

To **lower oil pressure**, decrease the number of shims behind the spring. This will lower the setting at which the valve opens.

After checking the parts and adjusting the number of shims, assemble the parts and install the locking nut (see Fig. 15). Then install the oil filter.

Fig. 15 — Pressure Regulating Valve Parts

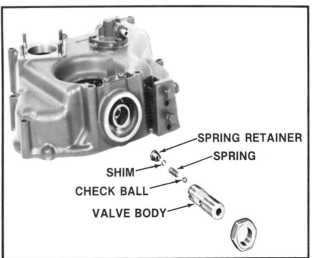

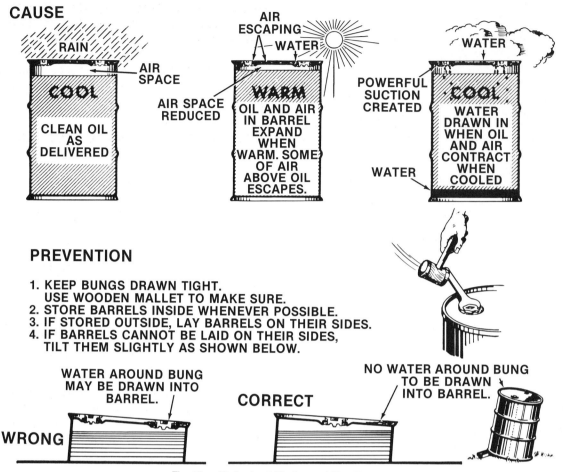

CAUSE

RAIN

AIR SPACE

COOL

CLEAN OIL AS DELIVERED

AIR SPACE REDUCED

AIR ESCAPING

WATER

WARM

OIL AND AIR IN BARREL EXPAND WHEN WARM. SOME OF AIR ABOVE OIL ESCAPES.

WATER

POWERFUL SUCTION CREATED

WATER

COOL

WATER DRAWN IN WHEN OIL AND AIR CONTRACT WHEN COOLED

PREVENTION

1. KEEP BUNGS DRAWN TIGHT.
   USE WOODEN MALLET TO MAKE SURE.
2. STORE BARRELS INSIDE WHENEVER POSSIBLE.
3. IF STORED OUTSIDE, LAY BARRELS ON THEIR SIDES.
4. IF BARRELS CANNOT BE LAID ON THEIR SIDES,
   TILT THEM SLIGHTLY AS SHOWN BELOW.

WATER AROUND BUNG MAY BE DRAWN INTO BARREL.

WRONG

CORRECT

NO WATER AROUND BUNG TO BE DRAWN INTO BARREL.

WATER

Fig. 16 — Methods of Storing and Handling Oil

CONTAMINATION OF OIL

Oil contamination will reduce engine life more than any other factor. Some sources of contamination are obvious while others are not.

Some of these sources of contamination and what can be done about them are as follows:

• The most obvious source of contamination is the **storing and handling of the oil** itself (see Fig. 16). If at all possible, store lubricants in a clean, enclosed storage area. Keep all covers and spouts in containers when not in use. These practices not only keep dirt out of the oil, but also reduce condensation of water caused by atmospheric changes.

• Another obvious source of contamination is **dust that is breathed into the engine** with intake air. It is very important that the air cleaner be cleaned or replaced regularly. At the same time, clean or replace the breather on the oil filler.

• A major source of contamination is due to **cold engine operation.** When the engine is cold, its fuel-burning efficiency is greatly reduced. Partially burned fuel may blow by the piston rings and into the crankcase. Oxidation of this fuel in the oil forms a very harmful varnish which collects on engine parts (Fig. 17). An overchoked or misfiring engine will also create this contamination from unburned fuel.

Water created by a cold engine contaminates the oil too. Water vapor, a normal product of combustion, tends to condense on cold cylinder walls. This condensation is also blown by the rings into the crankcase. The engine must warm up before this condensation problem is eliminated.

Water not only causes rusting of steel and iron surfaces, but it can combine with oxidized oil and carbon to form sludge. This sludge can effectively plug oil screens and passages.

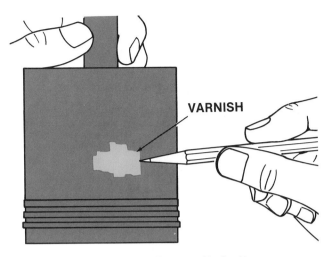

Fig. 17 — Varnish Forms on Engine Parts

Contamination from operating a cold engine can be prevented by:

A. *Warming up the engine before applying a load*

B. *Making sure the engine is brought up to operating temperature each time it is used*

C. *Using the proper thermostat to warm up the engine as quickly as possible*

• **Antifreeze** can be another sludge forming source of contamination. To guard against antifreeze contamination:

A. *Tighten torque head bolts to the specified torque during an overhaul*

B. *Use a cooling system sealer when filling the cooling system*

C. *Guard against detonation (gasoline engine) and improper use of starting fluids in diesel engine (both can result in cylinder head gasket damage).*

Other problems which cause oil contamination are:

• **Oxidation** is not an obvious source of contamination, but it is a very real one. Oxidation occurs when the hydrocarbons in the oil combine with oxygen in the air to produce organic acids. Besides being highly corrosive, these acids create harmful sludges and varnish deposits.

• **Carbon** particles are another contaminant created by the normal operation of the engine. The particles are created when oil around the upper cylinder walls is burned during combustion. Excessive deposits can cause the piston rings to stick in their grooves. Excessive carbon can also thicken oil to a point where it will not flow freely.

• **Engine wear** also creates a contaminant. Tiny metal particles are constantly being worn off bearings and other parts. These particles tend to oxidize and deteriorate the oil.

All sources of contamination cannot be eliminated, but there are several ways to protect the engine against oil contamination.

PROTECTING THE ENGINE AGAINST OIL CONTAMINATION

Start by using a good quality oil which has additives. Additives are put in the oil for a specific reason, based on the service expected for the oil.

Here are some of the important **additives:**

ANTICORROSION ADDITIVES protect metal surfaces from corrosive attack. These work with oxidation inhibitors.

OXIDATION INHIBITOR ADDITIVES keep oil from oxidizing even at high temperatures. They prevent the oil from absorbing oxygen, thereby preventing varnish and sludge formations.

ANTIRUST ADDITIVES prevent rusting of metal parts during storage periods, downtime, or even overnight. They form a protective coating which repels water droplets and protects the metal. These additives also help to neutralize harmful acids.

DETERGENT ADDITIVES help keep metal surfaces clean and prevent deposits. Particles of carbon and oxidized oil are held suspended in the oil. The suspended contaminants are then removed from the system when the oil is drained. Black oil is evidence that the oil is helping to keep the engine clean by carrying the particles in the oil rather than letting them accumulate as sludge.

However, remember that **additives eventually wear out.** To prevent this, drain the oil before the additives are completely depleted. Change the oil at intervals recommended by the manufacturer. Also service all the filters at regular intervals.

Finally, keep the fuel, cooling, and ignition systems in good condition so that the fuel is efficiently burned.

OIL CONSUMPTION

Some oil consumption is natural during normal operation of internal combustion engines.

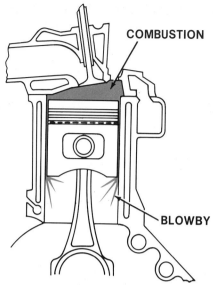

Fig. 18 — Engine Blowby

As we have already learned, lubricating oil provides a seal between the piston rings and cylinders. It is only natural that during the combustion process some of this sealing oil is burned up.

Excessive oil consumption can be caused by several conditions.

To properly diagnose the problem, follow this procedure:

1. First be sure the correct viscosity and grade of oil for the type of service and climate is being used. Oil that is too thin may "flood" the piston rings, while the oil that is too thick may "starve" the rings.

2. Be sure the engine has run long enough **under load** to insure that the rings have had a chance to seat. Some variation in oil consumption can be expected during break in, but it should be stabilized before 250 hours of operation. If not, assume that a problem exists and correct it at once.

3. Check the engine oil pressure and, if necessary, adjust the pressure regulating valve. High engine oil pressure can cause oil consumption by flooding the rings and valves with oil. Also check the crankcase breather. A plugged breather can increase the crankcase oil pressure.

4. Check for external oil leaks. What appears to be a small oil leak can add up to a considerable loss of oil. Drops of oil lost externally can add up to quarts of oil between oil change periods. Check the front and rear oil seals, all gaskets, and the filter attaching points.

5. Check for engine blowby (Fig. 18). Watch the fumes expelled through the crankcase vent tube. Fumes should be barely visible with the engine at fast idle with no load. If possible, compare the blowby with that of identical engines. This can be a guide to determine if the blowby is excessive.

Excessive blowby indicates that piston rings and cylinder liners have worn to the point where the rings cannot seal off the combustion chambers. An overhaul with new rings is then required. New or reconditioned pistons and liners may also be necessary.

6. If blowby is not excessive, check to see if the oil is being lost through the valve guides. A supply of oil is usually maintained in the rocker arm area to lubricate the rocker arms, valve stems, and valve guides.

Gravity, inertia, vacuum, and an atomizer effect all combine to force oil down between the valve stem and guide (Fig. 19).

To check for this condition, start the engine and warm it up. Let it idle slowly for ten minutes. Then remove the cylinder head, and inspect the valve ports and the undersides of the valve heads. If these are wet with oil, the oil has been drawn through the valve guides.

At the same time, check the piston heads. If they are wet, the oil has been drawn past the piston rings. If excessive blowby has indicated worn rings, this procedure can be used to confirm it.

Often oil consumption problems are blamed on worn rings when the real cause is worn valve guides. If the rings are replaced and not the valve guides, compression in the cylinder increases and forces even more oil through the valve guide.

7. Worn crankshaft bearings can also contribute to oil consumption. When worn, the bearings throw an excess of oil onto the cylinder walls. The action of the piston forces part of this excess oil into the combustion chambers where it is burned.

8. Excessive engine speeds are another common cause of oil consumption. Observe the fast idle limits set by the manufacturer.

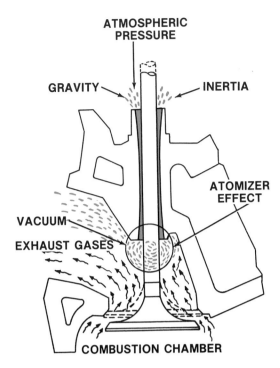

ATMOSPHERIC PRESSURE

GRAVITY

INERTIA

ATOMIZER EFFECT

VACUUM

EXHAUST GASES

COMBUSTION CHAMBER

Fig. 19 — Oil Loss Through Valve Guides

CHANGING ENGINE OIL

Oil loses many of its good lubricating qualities as it gets dirty and its additives wear out.

Acid formations, sludge, varnish, and engine deposits contaminate oil and make it unfit for continued use.

On the other hand, just because crankcase oil is black, doesn't mean it's time for an oil change.

Additives in the oil are supposed to clean and hold deposits in suspension for removal when the oil is drained.

Change the oil sometime **before** the additives wear out which protect the engine against sludge. For the average person, this time is practically impossible to determine. The best policy is to follow the manufacturer's recommendations on oil and filter changes.

New or rebuilt engines require oil and filter changes after a specified break in period. Performing this service on time is very important since foreign materials accumulate in the oil at a faster rate during initial operation than later when the engine is broken in.

When changing the oil and filter on any engine, always warm up the engine first. This way the contaminants and foreign materials are mixed with the oil and are drained out with it.

Replacing some oil filters requires installing new gaskets or sealing rings. Be sure the sealing surfaces on the engine and filter are clean.

After installing the filter and filling the engine with oil, run the engine and check for possible filter leaks.

Keep a record of all oil and filter changes to be sure of regular engine service.

CHAPTER 9 REVIEW

1. (Fill in the blanks) The two basic types of lubrication systems on 4-cycle engines are _____ _____ and _____ _____.

2. (Multiple choice) Of the following, which type of oil pump is **not** commonly found in pressure lubrication systems on 4-cycle engines:

   A. External-gear pump
   B. Rotor pump
   C. Plunger pump
   D. Vane pump

3. (True or false) Most wear in an external-gear pump will occur between the side of the gears and the housing.

4. (Fill in the blank) Generally, to raise oil pressure _____ shims behind the spring in the pressure regulating valve.

5. List at least five of the seven major sources or causes of oil contamination.

6. (Fill in the blank) The substances that are incorporated into engine oil to help the oil do a better job of protecting engine components is called _____.

7. (True or false) Excessive blowby indicates that piston rings and cylinder liners have worn to the point where the rings cannot seal off the combustion chamber.

8. (True or false) When crankcase oil turns black, it is time for an oil change.

CHAPTER 10

DIAGNOSIS AND TESTING OF ENGINES

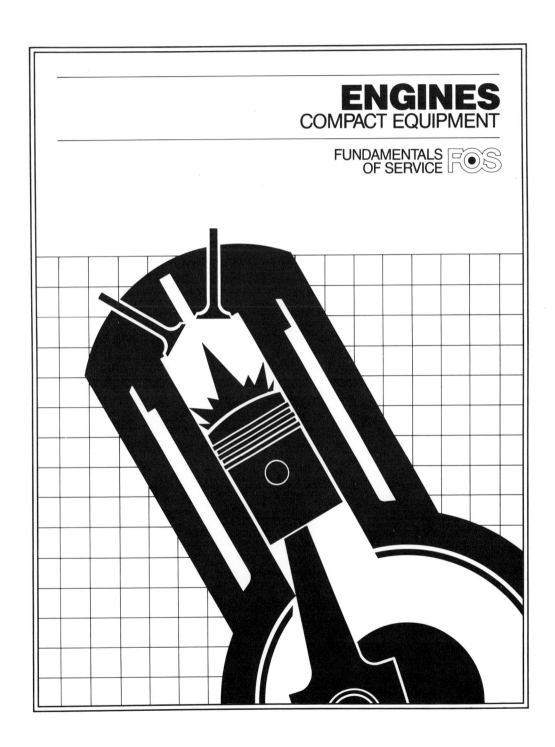

ENGINES
COMPACT EQUIPMENT

FUNDAMENTALS
OF SERVICE FOS

SKILLS AND KNOWLEDGE

This chapter contains basic information that will help you gain the necessary subject knowledge required of a service technician. With application of this knowledge and hands-on practice, you should learn the following:

• The basic diagnostic skills required of a good service technician

• The seven basic steps to good engine diagnosis

• Basic troubleshooting for gasoline and diesel fuel systems

• Basic troubleshooting for 2-cycle and 4-cycle engines

• How to make various tests of engines and fuel systems to locate engine problems

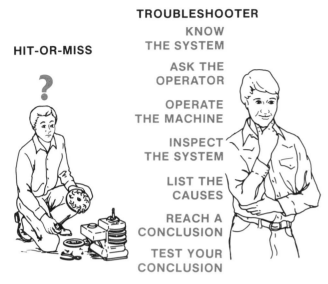

TROUBLESHOOTER

KNOW
THE SYSTEM

ASK THE
OPERATOR

OPERATE
THE MACHINE

INSPECT
THE SYSTEM

LIST THE
CAUSES

REACH A
CONCLUSION

TEST YOUR
CONCLUSION

HIT-OR-MISS

Fig. 1 — Which Would You Rather Be?

INTRODUCTION

For an engine to operate most efficiently, routine maintenance, or service, is required.

If engine failure occurs, however, the service technician must properly analyze the failure so that the correct repairs can be made. This manual does not attempt to provide an in-depth failure analysis. It does, however, provide basic guidelines to help the technician identify possible causes of failed parts.

After an engine problem is diagnosed, the repair is made. Repair includes disassembly, parts inspection, parts reconditioning or replacement, and assembly. During and after assembly, the engine must be properly adjusted for proper operation. Most of these repairs or adjustments have been discussed in previous chapters.

Many engine problems are electrical system problems. Electrical system diagnosis and testing is beyond the scope of this text but is covered in *FOS Electrical Systems For Compact Equipment*. See the Appendix for other Suggested Readings.

Hit-or-Miss or Professional technician — which would you rather be? Both have the title of technician but don't be fooled by that (Fig. 1).

The Hit-or-Miss technician is a parts exchanger who dives into an engine and starts replacing parts helter-skelter until the trouble is found — maybe — after wasting a lot of the customer's time and money.

The Professional technician starts out by using careful thought. All the facts are obtained and examined until the trouble has been pinpointed. Then the diagnosis is checked out by testing it and only then are parts replaced.

The Hit-or-Miss technician is fast becoming a thing of the past. What dealer can afford to keep such a technician around at today's prices?

With the complex systems of today, diagnosis and testing by the professional technician is the only way.

This chapter contains three basic sections to help guide the technician in diagnosing and testing engine problems. These sections are:

• **Basic diagnosis**

• **Troubleshooting guides**

• **Testing**

The first section will deal with the basic steps required for thorough engine diagnosis. The troubleshooting guides are more specific and help locate particular engine problems.

BASIC DIAGNOSIS — SEVEN BASIC STEPS

A good program of diagnosis and testing has seven basic steps:

1. **Know the system**

2. **Ask the operator**

3. **Inspect the engine**

4. **Operate the engine**

5. **List the possible causes**

6. **Reach a conclusion**

7. **Test your conclusion**

Let's see what these steps mean.

Fig. 2 — Know the System

HOW?
WHAT?
WHEN?
WHERE?

Fig. 3 — Ask the Operator

1. KNOW THE SYSTEM

In other words, "Do your homework." Study the engine Technical Manuals (Fig. 2). Know how the engine works, how it can fail, and the three basic needs — *fuel-air mixture, compression,* and *ignition.*

Keep up with the service bulletins. Read them and then file in the proper place. The solution to your problem may be in this month's bulletin.

You can be prepared for any problem by knowing the engine.

2. ASK THE OPERATOR

A good reporter gets the full story from a witness — the operator.

He can tell you how the engine acted when it started to fail; what was unusual about it.

What work was the engine doing when the trouble occurred? Was the trouble erratic or constant?

What did the operator do after the trouble? Did he attempt to repair it himself?

Ask how the engine is used and when it was serviced (Fig. 3). Many problems can be traced to poor maintenance or abuse of the engine.

Be tactful but get the full story from the operator.

3. INSPECT THE ENGINE

Go over the engine and inspect for all the items listed below.

Use your eyes, ears, and nose to spot any "tips" that lead to the source of trouble.

• Look for **water leaks** at the radiator, water pump, hoses, and around the cylinder head gasket.

• Check for **oil leaks** at the oil pan, drain plugs, and gaskets. Also inspect round crankshaft bores, oil seals, and valve cover. Look inside the flywheel housing for signs of oil.

• Look for **fuel leaks** at the tank, lines, filters, and pumps. Also check for restrictions or evidence of water in the fuel.

• Inspect for **clutch problems** which might affect the engine. Free travel should be adequate.

• Look for **other trouble signs** which could lead to future problems if not corrected early.

In general, look for anything unusual. Keep a list of all the trouble signs.

If the engine can be run, start it and warm it up. Then run it through its paces. Don't completely trust the operator's story — check it yourself.

Fig. 4 — Operate the Engine

4. OPERATE THE ENGINE

Next test engine operation. Before running the engine:

• Be sure that running the engine will not make the existing problem worse.

• Make sure the engine is clean. A thorough steam cleaning takes only minutes and removes dirt that can conceal obvious leaks and other faults.

• Make sure fluids are at the proper levels. Running an engine at low oil or coolant levels can aggravate engine problems.

• Do not run the engine if metal particles, water, or coolant were found in the oil.

Start the engine according to recommended procedures and warm it up thoroughly before making any tests (Fig. 4).

During operation take note of:

• **Starting response** — Long, slow cranking or heavy smoke from exhaust.

• **Gauge readings** — Check that readings are within normal ranges. Many types of compact equipment have no gauges or provisions for them.

• **Abnormal engine noises** — If a rattling noise is present, shut off the engine immediately. Loose or broken parts may be inside the engine. Note abnormal noises and the location as well as the conditions under which they occur, such as if noises occur only at specific engine speeds or loads or only when engine is warm or cold.

• **Exhaust smoke** — Check for sooty or oily exhaust, exhaust smoke coming from the cylinder head or gasket, strange odor (coolant being burned with the fuel), and heavy smoke or oil coming from the crank-case breather.

• **Controls** — Check for smooth operation and normal engine response to the control.

• **Engine response** — Observe engine for proper idle and for proper operation under load.

• **Governor response** — Check response as well as full load and full throttle idle rpm.

Compare your notes to the descriptions given by the operator.

Unusual gauge readings could mean the gauge is malfunctioning. Check the gauge by turning off the engine and removing the gauge line or sending unit for the suspect gauge. Replace it with a service department test gauge. Restart the engine and check the new readings. If the readings are now in the proper range replace the gauge or sending unit. If the measurements agree with those taken earlier, the problem is in the system being tested and will require further diagnosis.

Use your common sense to find out how the engine is operating.

Fig. 5 — List the Possible Causes

Fig. 6 — Test Your Conclusion

5. LIST THE POSSIBLE CAUSES

Now you are ready to make a list of the possible causes of the engine's troubles (Fig. 5).

What were the signs you discovered while inspecting and operating the engine?

- *Did the engine lack power?*
- *Any smoke from the crankcase vent?*
- *Did the engine run too hot or too cold?*
- *How was the oil pressure?*

Which of the signs tell you the most likely cause, which is second, etc.?

Are there any other possibilities? (One failure often leads to another).

6. REACH A CONCLUSION

Look over your list of possible causes and decide which are most likely and which are easiest to verify.

Use the troubleshooting guides shown later in this chapter.

Reach your decision on the leading causes and plan to check them first — after making the easy checks.

7. TEST YOUR CONCLUSION

Before you start repairing the system, test your conclusions to see if they are correct (Fig. 6).

Many of the items on your list can be verified without further testing.

Maybe you can isolate the problem to one system of the engine — lubrication, cooling, etc.

But the location within the system may be harder to find. This is where testing tools can help.

Later this chapter will tell how to test the engine and locate failures.

But first let's repeat the seven rules for good troubleshooting:

1. **Know the system**
2. **Ask the operator**
3. **Inspect the engine**
4. **Operate the engine**
5. **List the possible causes**
6. **Reach a conclusion**
7. **Test your conclusion**

Study the troubleshooting section that follows so you are familiar with typical problems and causes.

SUMMARY — DIAGNOSING ENGINE PROBLEMS

After the preceding diagnostic procedure is completed, evaluate all the information gathered up to this point. Determine if the symptoms indicate a specific problem or if further engine inspection for several possible problems is required (Fig. 7).

Fig. 7 — Requirements for Good Diagnosis

GOOD DIAGNOSIS

**What the Customer Wants
From Today's Technician**

- That only NECESSARY work be done
- That cost be REASONABLE
- That work be completed QUICKLY
- That problems be fixed RIGHT — THE FIRST TIME

The Only Way To Satisfy This Customer

- Quick, complete, and accurate DIAGNOSIS

And The Way To Good Diagnosis

- Discover the problem with the customer's help
- Write a clear and accurate repair order
- Test the diagnosis in a logical sequence

If a more thorough inspection requires engine disassembly, check the suspected problems that are easiest to fix first. For example, it may not be necessary to disassemble the engine if only a new cylinder head gasket will fix the problem.

TROUBLESHOOTING GUIDES

Use the troubleshooting guides on the following pages to help in listing all the possible causes of trouble when you diagnose and test an engine.

Once you have located the possible causes, reach a conclusion that you can test.

Use the engine technical manual for specific tests.

The troubleshooting guides are separated into gasoline engines and diesel engines. Each guide contains a table for quick reference to possible causes. The possible causes are listed in detail after each table.

TABLE 1
GASOLINE ENGINE TROUBLESHOOTING GUIDE
(2-Cycle and 4-Cycle Engines)

PROBLEM	FUEL SYSTEM CAUSES			IGNITION SYSTEM CAUSES		OTHER CAUSES							
	No Fuel	Improper Fuel	Improper Fuel-Air Mixture	No Spark	Poor Ignition	Insufficient Cooling	Improper Lubrication	Low Compression	Valve Problems (4-Cycle)	Carbon Build-Up	Faulty Governor	Engine Overloaded	Cold Engine
Will Not Start	X			X				X	X				
Hard Starting		X	X		X			X	X				
Engine Stops Running	X			X									
Lacks Power		X	X		X	X		X	X	X		X	X
Runs Erratically		X	X		X						X		
Knocks or Pings		X	X			X				X		X	
Skips or Misfires			X		X								
Backfires			X		X				X				
Overheats			X		X	X	X		X			X	
Idles Poorly			X		X								X
Engine Quits Suddenly	X			X			X		X				
Poor Acceleration			X		X	X		X	X	X	X	X	X

POSSIBLE CAUSES OF GASOLINE FUEL SYSTEM PROBLEMS (See Table 1)

No Fuel

1. Fuel tank empty
2. Fuel valve shut off
3. Filler cap vent plugged
4. Filter or lines clogged
5. Fuel pump faulty

Improper Fuel

1. Stale fuel
2. Water in fuel
3. Fuel octane too low

Improper Fuel Mixture

1. Overchoking — flooded
2. Air intake restricted
3. Carburetor set wrong
4. Intake manifold gaskets leaking
5. Fuel pump weak
6. Carburetor faulty
7. Reed plate faulty (2-cycle)
8. Crankcase air leak (2-cycle)

IGNITION SYSTEM CAUSES

No Spark

1. Ignition switch off or faulty
2. Leads disconnected or broken
3. Spark plug bad
4. Breaker points stuck or oxidized
5. Coil or condenser faulty

Poor Ignition

1. Spark plug gap excessive
2. Spark plug wet or fouled
3. Wrong spark plug
4. Cracked spark plug porcelain
5. Timing Wrong
6. Weak condenser or coil
7. Improper air gap (magneto)
8. Spark plug lead faulty insulation

OTHER CAUSES

Insufficient Cooling (air cooled)

1. Air intake screen plugged
2. Engine cooling fin air passages plugged

Insufficient Cooling (liquid cooled)

1. Low coolant level
2. Radiator plugged
3. Thermostat stuck
4. Loose or broken fan belt
5. Faulty water pump
6. Damaged fan blades

Improper Lubrication

1. Crankcase low on oil (4-cycle)
2. Incorrect oil
3. Oil screen plugged (4-cycle)
4. Oil pump faulty (4-cycle)
5. Oil galley plugged (4-cycle)
6. Crankshaft sleeve bearing oil hole not aligned with oil galley hole (4-cycle pressure oil system)
7. Air leaks in crankcase or defective oil seals
8. Faulty oil pressure system (4-cycle)
9. Incorrect fuel-oil mixture (2-cycle)

Low Compression

1. Leaking cylinder head gaskets
2. Warped cylinder head
3. Improper valve clearance (4-cycle)
4. Burned or warped valves (4-cycle)
5. Broken valve springs (4-cycle)
6. Worn piston rings
7. Worn cylinder bore
8. Damaged piston
9. Warped reed valves (2-cycle)

Valve Problems (4-Cycle)

1. Valve clearance not correct
2. Valves not seating properly
3. Burned valve
4. Stuck valve
5. Broken valve spring
6. Keepers improperly installed or broken valve stem
7. Valves not opening and closing properly

Carbon Build-Up

1. Carbon on cylinder head, valves (ports) and piston

Faulty Governor

1. Improper governor sensitivity adjustment
2. Fault governor

Engine Overloaded

1. Working too fast for engine
2. Applying excessive load

Cold Engine

1. Fuel not burning completely
2. Defective thermostat (liquid-cooled engine)

TABLE 2
DIESEL ENGINE FUEL SYSTEM
TROUBLESHOOTING GUIDE
(See Table 3 for other diesel engine problems)

PROBLEM	No Fuel	Improper Fuel	Air in Fuel System	Fuel Filter Clogged	Faulty Nozzles	Faulty Injection Pump
Engine:						
Will not start	X	X	X	X	X	X
Hard starting		X	X	X	X	X
Stops running	X	X	X	X		X
Lacks power		X	X	X	X	X
Runs erratically		X	X	X	X	X
Knocks					X	X
Misfires		X	X		X	X
Overheats		X			X	X
Idles poorly					X	X
Quits suddenly	X	X	X	X		X
Poor acceleration		X		X	X	X
Too much black smoke					X	X

POSSIBLE CAUSES OF DIESEL FUEL SYSTEM PROBLEMS (See Table 2)

No Fuel

1. Fuel tank empty
2. Fuel valve shut off
3. Filler cap vent plugged
4. Filter or lines clogged
5. Fuel pump faulty
6. Air in fuel system

Improper Fuel

1. Fuel cetane too low
2. Water in fuel
3. Not winter fuel blend (cold temperature operation)

Air In Fuel System

1. Air in fuel pump
2. Air in fuel lines

Fuel Filter Clogged

1. Dirt in filter
2. Not winter fuel blend (cold temperature operation)

Faulty Nozzles

1. Nozzle orifices plugged or damaged parts
2. Broken nozzle valve spring
3. Cracked or split nozzle tip or body
4. Internal nozzle leak
5. Worn nozzle valve seat

Faulty Injection Pump

1. Incorrect fast idle speed
2. Incorrect pump timing
3. Nozzle return lines clogged

TABLE 3
DIESEL ENGINE TROUBLESHOOTING GUIDE
(See Table 2 for fuel problems)

PROBLEM	Low Compression	Valve Problems	Insufficient Cooling	Improper Lubrication	Cold Engine	Faulty Governor	Engine Overload	Air Intake Problem	Turbocharger Faulty
Engine:									
Will not start	X	X							
Hard starting	X	X							
Lacks power	X	X	X			X	X	X	X
Runs erratically						X			
Knocks		X						X	
Misfires						X			
Overheats			X	X			X		
Idles poorly	X	X			X			X	
Quits suddenly			X	X					
Poor acceleration	X	X				X	X	X	X
Too much white smoke	X				X				
Too much black smoke							X	X	X

POSSIBLE CAUSES OF DIESEL ENGINE PROBLEMS (See Table 3)

Low Compression

1. Leaking cylinder head gaskets
2. Warped cylinder head
3. Improper valve clearance
4. Burned or warped valves
5. Broken valve springs
6. Worn piston rings
7. Worn cylinder bore
8. Damaged piston
9. Faulty precombustion

Valve Problems

1. Valve clearance not correct
2. Valves not seating properly
3. Burned valve
4. Stuck valve
5. Broken valve spring
6. Keepers improperly installed or broken valve stem
7. Valves not opening and closing properly

Insufficient Cooling

1. Low coolant level
2. Radiator plugged
3. Thermostat stuck
4. Loose or broken fan belt
5. Faulty water pump
6. Damaged fan blades

Improper Lubrication

1. Crankcase low on oil (4-cycle)
2. Incorrect oil
3. Oil screen plugged (4-cycle)
4. Oil pump faulty (4-cycle)
5. Oil galley plugged (4-cycle)
6. Crankshaft sleeve bearing oil hole not aligned with oil galley hole (4-cycle pressure oil system)
7. Air leaks in crankcase or defective oil seals
8. Faulty oil pressure system (4-cycle)

Cold Engine

1. Engine not allowed to warm up
2. Faulty cooling system thermostat

Faulty Governor

1. Improper governor sensitivity adjustment
2. Faulty governor

Engine Overload

1. Ground speed too fast
2. Applying excessive load

Air Intake Problem

1. Air leak in intake system
2. Air cleaner restricted

Turbocharger Faulty

1. Air leak in intake or exhaust manifold
2. Improper clearance of turbine wheel
3. Bearings not lubricated
4. Broken turbine blades
5. Clogged intake manifold
6. Dirt buildup in compressor

TESTING THE ENGINE

Once you have reached a conclusion on the cause of the problem, you will have to test your conclusion by either inspection or an actual test procedure. The following discussion describes some of the typical tests performed on engines.

Tests other than these may be performed on engines, but in any case, refer to the technical manual for proper procedures.

The test procedures covered in this section are broken down into two basic categories with specific tests listed under each.

CATEGORY 1 — GENERAL ENGINE

- **Compression test**
- **Manifold pressure test**
- **Crankcase pressure test**
- **Oil pressure test**
- **Cooling system test**
- **Engine power (dynamometer) test**

CATEGORY 2 — FUEL SYSTEM

- **Carburetor**
- **Reed valve**
- **Diesel fuel injection pump**
- **Diesel fuel injection nozzle**

CATEGORY 1 - GENERAL ENGINE

TESTING COMPRESSION PRESSURE

Low compression pressure can result in hard starting, power loss and poor acceleration. Low compression can be caused by worn piston rings and cylinder wall, valve problems or cylinder leaks. By running a compression test, you can quickly tell if the problem is in the combustion area of the engine.

The following compression tests cover typical procedures for 2-cycle and 4-cycle gasoline engines and diesel engines.

Compression Test (Gasoline 2-Cycle)

TESTING COMPRESSION (SMALL 2-CYCLE ENGINES)

Disconnect spark plug lead to make sure engine will not start. Crank engine by hand in direction of normal rotation. This should be done when engine is cold. There should be considerable resistance as piston approaches top-dead-center. Hold piston against compression for several seconds. If resistance to pull decreases rapidly, it indicates poor compression. Poor compression is usually the result of worn piston rings, worn cylinder bore, or ring gaps not staggered around the piston.

TESTING COMPRESSION (LARGER 2-CYCLE ENGINES)

1. Remove the spark plug.

2. Install a compression gauge.

3. Pull recoil start rope several times or crank the engine with the starter (if it is electric start) to stabilize reading.

4. Compression pressure should be equal to or above the manufacturer's specifications.

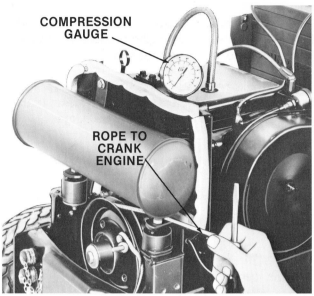

Fig. 8 — Install Compression Gauge

Compression Test (Gasoline 4-Cycle)

NOTE: Compression should be checked while the engine is warm. Run the engine for several minutes at full throttle. Shut the engine off and remove spark plug(s). Do not reinstall the spark plug(s) until you have checked compression.

Set throttle and choke in wide open position.

Place a compression gauge, Fig. 8, in the spark plug hole. Crank the engine for several seconds and observe the compression gauge reading. Check each cylinder several times.

Pressure readings that are equal to or above specifications indicate good compression. Low readings indicate a blown head gasket, warped cylinder head, worn piston rings, leaky valves, incorrect valve clearance or piston damage.

Compression Test (Diesel)

Make a compression test on a diesel engine by following these procedures.

NOTE: Before beginning test, insure that battery is fully charged.

IMPORTANT: Thoroughly clean external area around injection nozzles.

1. Remove fuel injection and fuel leak-off lines from injection nozzles.

2. Remove nozzle retainers and nozzles.

3. Remove precombustion chambers (if so equipped) from each nozzle bore, except the bore on cylinder to be tested.

4. Install adapter (Fig. 9) in injection nozzle bore, and reinstall nozzle retainer.

5. Attach test gauge to adapter.

6. Remove wire from thermo-start unit (if so equipped).

7. Make sure throttle is in the idle position.

8. Make sure decompression device (if so equipped) is not engaged.

9. Crank engine at specified speed and check gauge reading. Repeat check for each cylinder, and compare reading taken to specifications.

Pressure given in the manual was taken at a given altitude. A reduction in gauge pressure will result for readings taken at higher altitudes. See the technical manual for specific altitudes and readings.

An engine with low compression pressures will be hard to start and is in need of overhaul.

If pressure is much lower than specified, remove the gauge and apply oil to the ring area of piston through

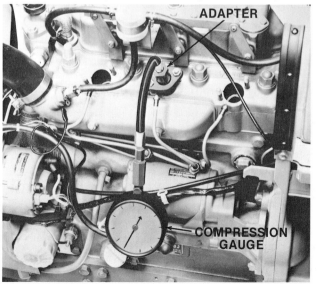

Fig. 9 — Installing Compression Gauge on Diesel Engine

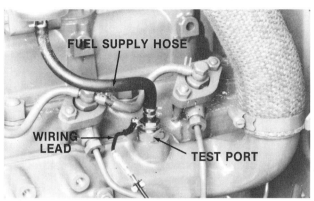

Fig. 10 — Manifold Intake Pressure Test Port

injection nozzle hole. Do not use too much oil and do not get oil on the valves.

Test compression again. A higher pressure reading indicates worn or stuck piston rings. If the pressure is still low, it is possible that valves are worn or sticking.

It is very important that all cylinder pressures be approximately alike. There should be less than about 25 psi (172 kPa) difference between cylinder pressures.

MANIFOLD INTAKE PRESSURE TESTING

Test manifold intake pressure at the designated test port on the manifold. This test is generally made on multiple-cylinder, 4-cycle engines. Before testing, disconnect any wiring lead, fuel supply hose, or starting assist plug from the test port (Fig. 10).

Next, install a pressure gauge to the test port (Fig. 11), start the engine, warm the engine to operating temperature, and check the gauge reading.

If the reading is not near the manufacturer's specification, check for a restricted air cleaner, loose intake hose connections, or a defective turbocharger.

CRANKCASE PRESSURE TESTING

The crankcase pressure test is used on 2-cycle gasoline engines. An improperly sealed crankcase can cause:

- **Hard starting**
- **Erratic operation**
- **Low power**
- **Overheating**

To test crankcase pressure, use the following procedure.

1. Remove the muffler.

2. Use the muffler gasket as a template to make a seal out of gasket paper material (Fig. 12).

3. Install the seal and muffler.

4. Remove the carburetor.

5. Use the carburetor spacer gasket as a template to make a seal out of gasket paper material.

6. Remove the spark plug and position the piston at bottom-dead-center.

Fig. 11 — Checking Manifold Intake Pressure

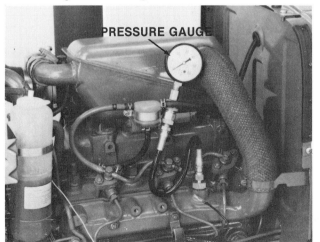

Fig. 12 — Seal for Checking Crankcase Pressure

Fig. 13 — Pump and Pressure Gauge Installed.

7. Install the pump and pressure gauge to the spark plug hole (Fig. 13).

8. Apply the specified pressure and check that the pressure loss does not exceed the specification.

9. If the pressure loss is excessive, use a soap solution to check crankcase seams, the cylinder base gasket, and the crankshaft seal for leakage.

10. Replace seals that are leaking.

OIL PRESSURE TESTING

Oil pressure tests are used only on pressure-lubricated, 4-cycle engines. To check oil pressure, install an oil pressure gauge to the test port (Fig. 14).

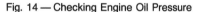

Fig. 14 — Checking Engine Oil Pressure

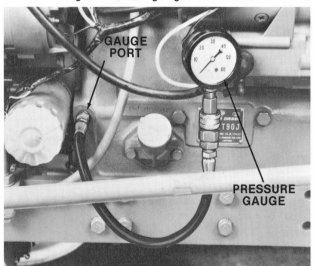

Start and run the engine at fast idle speed. When the engine is warmed to operating temperature, check the reading on the pressure gauge. If the reading does not fall within the limits specified by the manufacturer, remove the pressure regulating valve for service. This procedure was described in Chapter 9.

COOLING SYSTEM TESTING

Several cooling system tests may be used to check for engine cooling problems. They are:

- **Cooling system leak test**
- **Radiator cap test**
- **Thermostat test**
- **Exhaust gas leakage test**
- **Air-in-system test**

Each of these will be discussed separately.

Cooling System Leak Test

Most liquid-cooling systems are pressurized. Therefore, the radiator and the rest of the cooling system must be air tight for the system to operate properly. Before servicing the radiator, test the entire system for leaks.

Install a *pressure tester* on the filler neck of the radiator (Fig. 15). Pump the handle of the tester until the gauge reads the recommended system pressure. Carefully inspect radiator, water pump, hoses, drain cock, and cylinder block for leakage.

Mark each leak so it can easily be located when repairs are made.

Fig. 15 — Pressure Tester for Leak Testing Cooling System

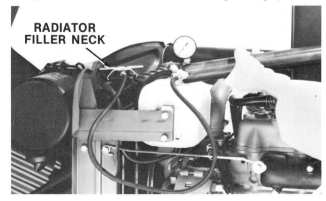

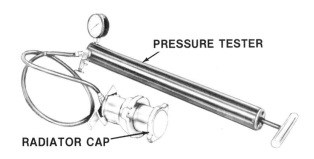

Fig. 16 — Pressure Test Radiator Cap

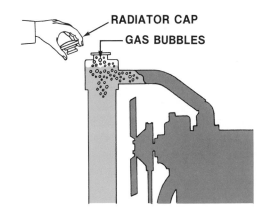

Fig. 18 — Check for Bubbles in Coolant

Radiator Cap Test

To test the radiator cap, connect it to the same pressure tester used to test the entire system (Fig. 16). The tester usually comes with a special adapter to accommodate the radiator cap.

Pump the handle of the pressure tester until the gauge indicates the recommended pressure. The gauge should hold at that pressure. If it doesn't, the radiator cap is damaged and must be replaced.

Thermostat Test

1. Suspend the thermostat and a thermometer in a container of water (Fig. 17). Do not let them rest against the side or bottom.

2. Heat and stir the water.

3. The thermostat should begin to open at the temperature stamped on it, plus or minus 10°F (5.5°C). It should be fully open — approximately 1/4 inch (6 mm) — at 22°F (12°C) above the stated temperature.

Fig. 17 — Testing the Thermostat

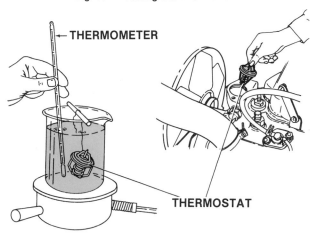

4. Remove thermostat and observe its closing action.

5. If the thermostat is defective, discard it.

Exhaust Gas Leakage Test

1. Warm up the engine and keep it under load.

2. Remove the radiator cap and look for excessive bubbles in the coolant (Fig. 18).

3. Either bubbling or an oil film on the coolant is a sign of blowby in the engine cylinders.

NOTE: Make this test quickly, before boiling starts, since steam bubbles give misleading results.

Air-In-System Test

If air leaks in the cooling system are suspected, the following checks can be made:

1. Adjust coolant to correct level.

2. Replace pressure cap with a plain, air tight cap.

3. Attach rubber tube to lower end of overflow pipe. Be sure radiator cap and tube are air tight.

4. With transmission in neutral gear, run engine at high speed until temperature gauge stops rising and remains stationary.

5. Without changing engine speed or temperature, put end of rubber tube in bottle of water.

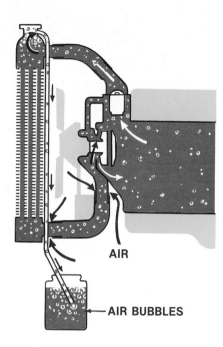

AIR

AIR BUBBLES

Fig. 19 — Check for Air in Cooling System

6. Watch for a continuous stream of bubbles in the water bottle, showing that air is being drawn into the cooling system (Fig. 19).

ENGINE POWER (DYNAMOMETER) TESTING

The dynamometer test is used for both 2- and 4-cycle engines.

If possible, test the engine on a dynamometer before it is tuned. This test gives the indicated PTO power output and fuel consumption of the engine as it is. This will help determine if a tune-up can restore the engine or whether an overhaul is needed.

Good performance by the engine depends on these basic things:

1. An adequate supply of clean air and fuel

2. Good compression

3. Proper valve and injection pump timing for good combustion (4-cycle diesel engine)

Make the dynamometer test as follows. The test can be made on 4-cycle and 2-cycle engines. The following procedure is for 4-cycle engines.

1. Connect the engine to the dynamometer using the manufacturer's instructions.

Fig. 20 — Testing for Carburetor Operation

2. Operate the engine at one-half load until the coolant and crankcase oil temperature are up to normal.

3. Run engine at fast idle.

4. Gradually increase the load on the engine until its speed is reduced to the specified load rpm.

5. Read the power on the dynamometer.

6. Compare the reading taken with the rated horsepower for the machine tested.

CATEGORY 2 — FUEL SYSTEM

Carburetor Testing

To make sure the carburetor is functioning, perform the following test.

1. Remove the spark plug and place your thumb over the spark plug hole (Fig. 20).

2. Pull the recoil starter rope several times or crank the engine with the starter (electric-start).

3. Your thumb should be moist with fuel.

Some diaphragm-type carburetors with a metering lever may be tested further. The procedures will be discussed next.

Diaphragm-Type Carburetor

The following test will quickly detect two things to indicate the condition of the carburetor or gasoline engines:

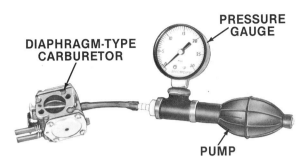

Fig. 21 — Testing One Type of Diaphragm-Type Carburetor

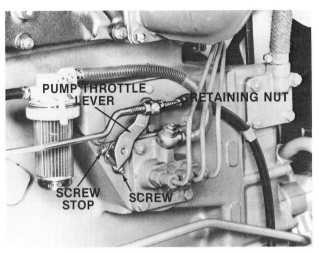

Fig. 23 — Disconnect Throttle Linkage

• *The condition of the inlet needle and seat and the metering lever spring.*

• *The sealing ability of the carburetor gaskets, diaphragms, and welch plug.*

Connect a pump with a pressure gauge as shown in Fig. 21 to make the tests described below. The carburetor must contain fuel to make the test.

TEST 1

Build up pressure to about 7 psi (50 kPa). The pressure should hold or reduce only slightly. If pressure leaks off quickly, check the fuel inlet needle seat and the position and condition of the metering lever spring.

TEST 2

Carefully seat the idle and high-speed mixture needles. Then build up pressure to about 15 psi (100 kPa). The pressure should hold or reduce only slightly. If pressure leaks off quickly, inspect gaskets, diaphragms, and welch plug for leakage.

REED VALVE TESTING

Reed valves on 2-cycle gasoline engines can be tested for leakage. To do this, remove the air cleaner from the carburetor intake and either start the engine or rapidly pull the starter rope several times.

Hold a piece of paper about one inch (25 mm) from the carburetor intake (Fig. 22). If fuel spots begin to appear on the paper, the reeds are leaking and must be replaced.

DIESEL INJECTION PUMP TESTING

The main tests required for the fuel injection pump is the **fast idle speed test** and the **pump gallery pressure test**.

Fast Idle Speed Test

To test pump fast idle speed, first start the engine and warm it to operating temperature. Next disconnect the throttle linkage from the injection pump throttle lever (Fig. 23).

Remove the PTO shaft cover and engage the PTO. Then use a string or wire to tie the pump throttle lever so that the screw (Fig. 23) is contacting the screw stop.

Fig. 22 — Testing for Reed Valve Leakage

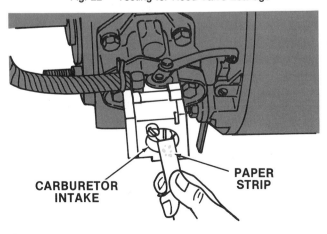

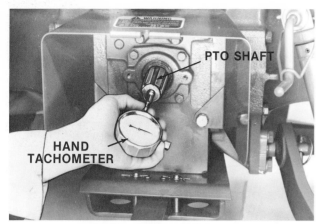

Fig. 24 — Measure PTO Speed

Fig. 26 — Install Pressure Gauge to Injection Pump

Use a hand tachometer (Fig. 24) to measure PTO shaft speed. If the speed does not match that specified in the technical manual, adjust the screw to obtain the correct speed.

After making the adjustment, install the PTO shaft cover and connect the throttle linkage back to the injection pump throttle lever.

Injection Pump Gallery Pressure Test

This test is used to determine if the filter is clogged or the transfer pump is faulty. To make this test, remove the bleed screw from the injection pump (Fig. 25).

Install a pressure gauge to this port (Fig. 26) and crank the engine. Note the reading on the gauge.

If pressure is lower than specified, replace the fuel filter and repeat the test. If pressure is still too low, replace the fuel check valve assembly (if so equipped) (Fig. 27). If the pressure is still too low, the pump must be repaired or adjusted on a test stand.

DIESEL INJECTION NOZZLE TESTING

The tests for injection nozzles are for pressure, leakage, chatter, and spray pattern.

CAUTION: Escaping fluid under pressure can penetrate the skin causing serious injury (Fig. 28, left). Relieve pressure before disconnecting fuel or other lines. Tighten all connections before applying pressure. Keep hands and body away from pinholes and nozzles which eject fluids under high pressure. Use a piece of cardboard or paper to search for leaks (Fig. 28, right). Do not use your hand.

If ANY fluid is injected into the skin, it must be surgically removed within a few hours by a doctor familiar with this type injury or gangrene may result.

OPENING PRESSURE TEST

Connect the nozzle and fuel line to a nozzle tester (Fig. 29). Place a container under the nozzle to catch fuel. Fill the nozzle tester with diesel fuel.

Fig. 25 — Remove Bleed Screw from Injection Pump

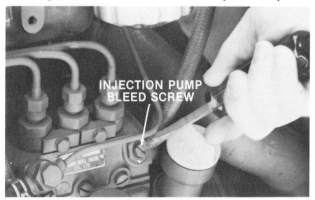

Fig. 27 — Fuel Check Valve Assembly

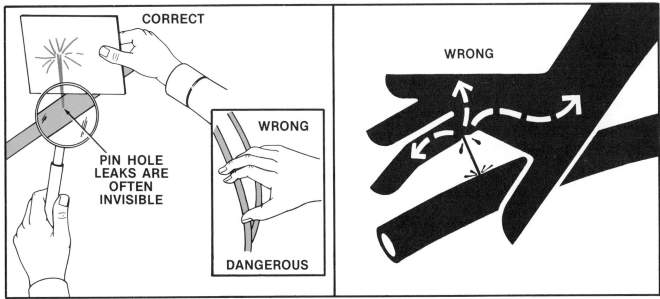

Fig. 28 — Fluid Under Pressure Can Cause Injury

IMPORTANT: Use clean, filtered fuel when testing the injection nozzle. Nozzle damage can result from using dirty fuel for testing.

Pump the handle of the nozzle tester several times to flush out the nozzle and fittings and to remove all air from the nozzle.

Pump the handle slowly and steadily to raise fuel pressure. Observe the reading on the pressure gauge as the nozzle begins to open. Recheck the nozzle opening pressure by releasing the pressure and repeating the test.

If the nozzle opening pressure is not that which is the recommended specification, turn the pressure adjusting screw on the nozzle (Fig. 30) to obtain the correct pressure. Then recheck the pressure and adjust the screw until it is within specification.

Leakage Test

To test the nozzle for leakage, again use the nozzle tester (see Fig. 29). Wipe the nozzle dry and slowly pump the handle until the pressure on the gauge shows slightly less than nozzle opening pressure.

Fig. 29 — Injection Nozzle Connected to Tester

Fig. 30 — Turning the Pressure Adjusting Cap

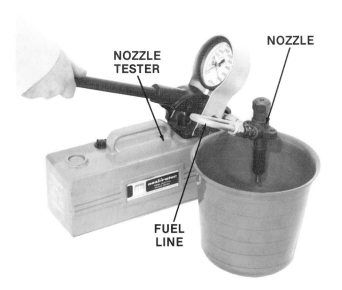

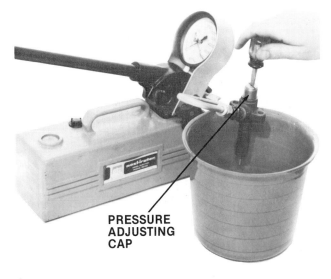

Watch for fuel to drip from the spray orifice. If fuel does drip within a few seconds, replace the nozzle.

Also check for leaks around the threaded connections.

Chatter And Spray Pattern Test

To test for nozzle chatter and correct spray pattern, again use the nozzle tester and operate the handle rapidly (four to six short strokes per second). A soft chatter in the nozzle is normal. Failure to chatter may be caused by a binding or bent nozzle valve.

NOTE: When the lever is moved fast enough to cause the nozzle to chatter, the spray pattern from the nozzle should be broad and finely atomized. During slow lever movement, test fuel will stream from the nozzle.

A partially clogged or eroded nozzle orifice will usually cause an irregular spray pattern. The spray may also be streaky rather than finely atomized.

Disassemble the nozzle for cleaning or repair if it does not chatter or spray properly.

SUMMARY

A good service technician thinks a problem through before beginning repairs. A technician who merely replaces parts until a problem is corrected is an expensive liability.

There are seven basic steps involved in a thorough engine diagnosis. The steps are:

1. **Know the system**

2. **Ask the operator**

3. **Inspect the engine**

4. **Operate the engine**

5. **List the possible causes**

6. **Reach a conclusion**

7. **Test your conclusion**

Engine problems may be related to the interior engine components or the fuel system. Charts are available to help the technician quickly identify possible causes of engine problems.

Various tests are available to further help identify engine problems. These tests may be related to the engine in general or to the fuel system.

General Engine Tests

- **Compression**
- **Manifold pressure**
- **Crankcase pressure**
- **Oil pressure**
- **Cooling system**
- **Engine power (dynamometer)**

Fuel System Tests

- **Carburetor**
- **Reed valves**
- **Fuel injection pump**
- **Fuel injection nozzles**

CHAPTER 10 REVIEW

1. List the seven basic steps to good engine diagnosis.

2. (Multiple choice) The **least** likely cause for an improper fuel-air mixture in a gasoline engine is?

   A. Air intake restricted
   B. Carburetor set wrong
   C. Carburetor faulty
   D. Filter or lines clogged
   E. Overchoking flooded

3. List five of the factors that could cause valve problems in a 4-cycle engine.

4. (Multiple choice) Low compression in a diesel engine is most likely caused by:

   A. Thermostat stuck
   B. Low oil pressure
   C. Improper valve clearance
   D. Engine not warmed to operating temperature

5. (True or false) Low compression pressure can result in hard starting, power loss, and poor acceleration.

6. Discuss some of the problems that an improperly sealed crankcase on a 2-cycle engine can cause.

7. (Fill in the blanks) _____ _____ leakage shows up as bubbles or an oil film on the coolant in the radiator after the engine is warmed up.

8. (True or false) Test reed valves on a 2-cycle by removing a spark plug, placing your thumb over the hole, and cranking the engine.

9. (Fill in the blanks) Two tests that may be made on diesel fuel injection pumps are _____ _____ speed test and _____ pressure test.

APPENDIX

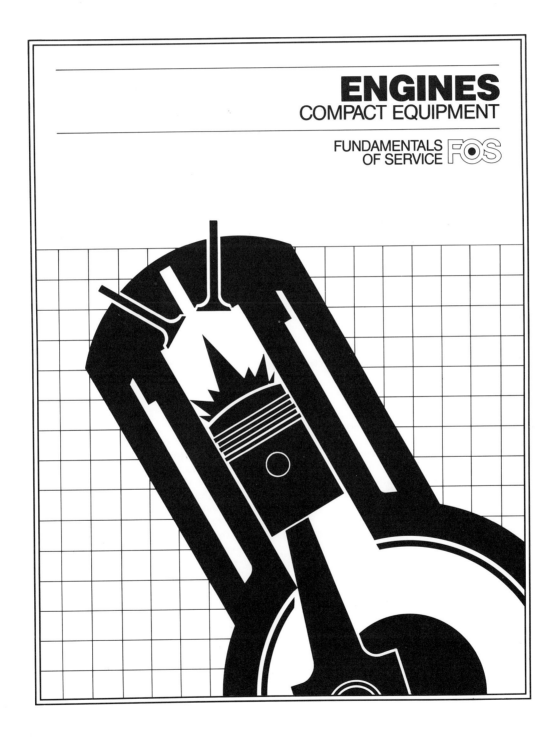

ENGINES
COMPACT EQUIPMENT

FUNDAMENTALS
OF SERVICE FOS

FLOAT-TYPE CARBURETOR

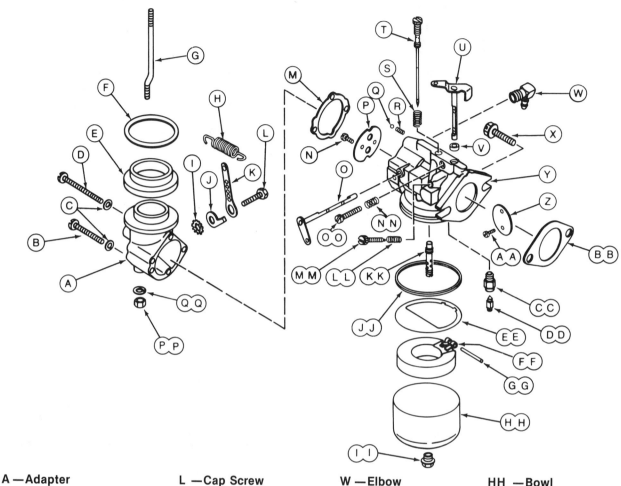

A —Adapter
B —Screw (2 used)
C —Washer (3 used)
D —Screw
E —Adapter Ring
F —Gasket
G —Stud
H —Governor Spring
I —Tooth Washer
J —Stop
K —Governor Speed Control
    Lever

L —Cap Screw
M —Gasket
N —Screw (2 used)
O —Choke Shaft
P —Choke Valve
Q —Ball Check Valve
R —Spring
S —Spring
T —High-Speed Mixture
    Needle
U —Throttle Shaft
V —Bushing

W —Elbow
X —Screw (2 used)
Y —Carburetor
Z —Throttle Valve
AA —Screw (2 used)
BB —Gasket
CC —Valve
DD —Seat
EE —Baffle
FF —Float
GG —Pin

HH —Bowl
II —Screw
JJ —Gasket
KK —Float Valve
LL —Spring
MM —Idle-Speed Mixture
     Screw
NN —Spring
OO —Idle-Speed Screw
PP —Stud Nut
QQ —Lock Washer

DIAPHRAGM-TYPE CARBURETOR

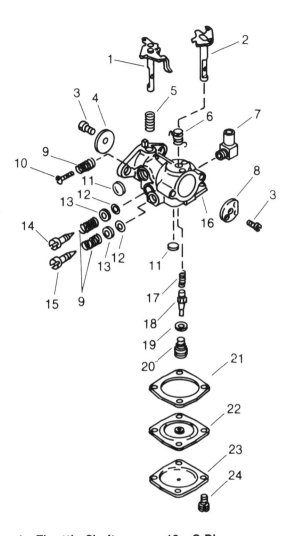

VACUUM-SUCTION-LIFT CARBURETOR

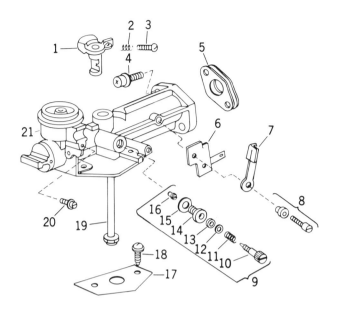

1—Throttle
2—Spring
3—Screw (No. 8 x 5/8")
4—Screw (2 used)
5—Gasket
6—Insulator Plate
7—Lever
8—Screw and Bushing
9—Mixture Adjusting Needle Kit
10—Mixture Adjusting Needle
11—Spring
12—Washer (2 used)
13—Packing
14—Nut
15—Gasket
16—Seat
17—Gasket
18—Screw (2 used)
19—Fuel Pipe
20—Screw and Washer
21—Body

1—Throttle Shaft
2—Choke Shaft
3—Machine Screw
4—Throttle Shutter
5—Throttle Return Spring
6—Choke Return Spring
7—Fuel Inlet Fitting
8—Choke Shutter
9—Spring
10—Idle Stop Screw
11—Welch Plug
12—O-Ring
13—O-Ring
14—Idle Mixture Needle
15—High-Speed Mixture Needle
16—Carburetor Body
17—Needle Valve Spring
18—Needle Valve
19—Valve Seat Gasket
20—Valve Seat
21—Diaphragm Gasket
22—Diaphragm
23—Diaphragm Cover
24—Screw

PULSATING-SUCTION-LIFT CARBURETOR

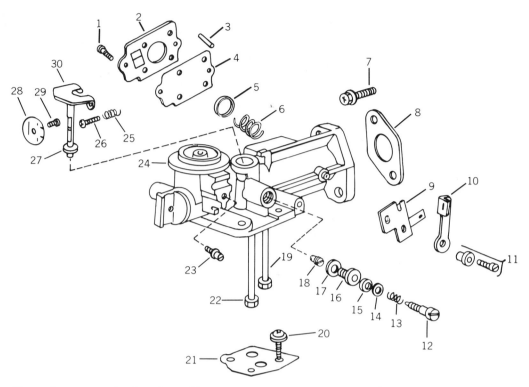

1—Screw (4 used)
2—Cover
3—Pin
4—Diaphragm
5—Cap
6—Spring
7—Screw (2 used)
8—Gasket

9—Insulator Plate
10—Lever
11—Screw and Bushing
12—Mixture Adjusting
    Needle
13—Spring
14—Washer (2 used)
15—Packing

16—Nut
17—Washer
18—Seat
19—Plastic Fuel Pipe
20—Screw (4 used)
21—Gasket
22—Brass Fuel Pipe Screen
23—Screw and Washer

24—Body
25—Spring
26—Screw (No. 5 x 1/2")
27—Felt Washer
28—Throttle Valve
29—Screw
30—Throttle Lever

TURBOCHARGER

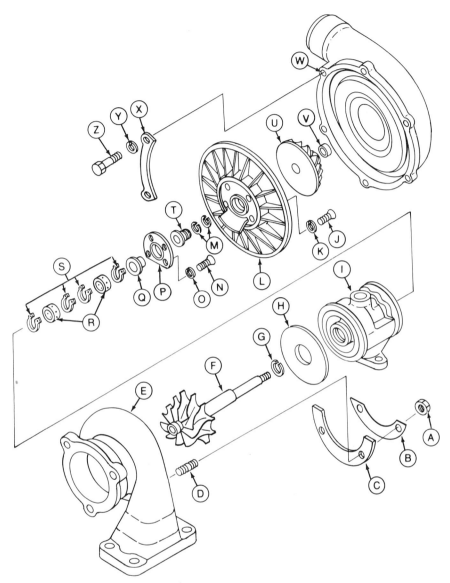

A —Nut (4 used)
B —Locking Plate (2 used)
C —Lock Plate (turbine side)
D —Stud Bolt (4 used)
E —Turbine Housing
F —Turbine Wheel and Shaft
G —Seal Ring (turbine side)
H —Heat Protector
I —Center Housing
J —Screw (4 used)
K —Toothed Lock Washer (4 used)
L —Seal Plate
M —Seal Ring (compressor side)

N —Screw (4 used)
O —Toothed Lock Washer (4 used)
P —Thrust Plate
Q —Thrust Bushing
R —Journal Bearing
S —Retaining Ring (4 used)
T —Oil Thrower
U —Compressor Wheel
V —Lock Nut
W—Compressor Housing
X —Lockplate (compressor side)
Y —Washer (6 used)
Z —Cap Screw (6 used)

SUGGESTED READINGS

FUNDAMENTALS OF SERVICE: *Electrical Systems — Compact Equipment*, John Deere Service Training, Dept. F., John Deere Road, Moline, IL 61265

FUNDAMENTALS OF SERVICE: *Hydraulics — Compact Equipment*, John Deere Service Training, Dept. F., John Deere Road, Moline, IL 61265

FUNDAMENTALS OF SERVICE: *Power Trains — Compact Equipment*, John Deere Service Training, Dept. F., John Deere Road, Moline, IL 61265

Small Engines — Volume 1, Care and Operation, American Association for Vocational Instructional Materials, Engineering Center, Athens, GA 30602

Small Engines — Volume 2, Maintenance and Repair, American Association for Vocational Instructional Materials, Engineering Center, Athens, GA 30602

Small Gas Engines, Roth, Alfred C. and Baird, Ronald J., The Goodheart — Wilcox Company, Inc., South Holland, IL 60473

GLOSSARY

A

ABRASION — Wearing or rubbing away of a part.

ADDITIVE — A substance added to oil to give it certain properties. For example, a material added to engine oil to lessen its tendency to congeal or thicken at low temperatures.

AIR CLEANER — A device for filtering, cleaning, and removing dust from the air drawn into an engine.

AIR COOLING — A method of removing heat from an engine by passing air across the engine to carry away the heat.

ANTIFREEZE — A material such as alcohol, glycerin, etc., added to water to lower its freezing point.

ANTIFRICTION BEARING — A bearing constructed with balls, rollers, or the like between the journal and the bearing surface to provide rolling instead of sliding friction.

ATMOSPHERIC PRESSURE — The weight of the air at sea level; about 14.7 lbs. per square inch (100 kPa); less at higher altitudes.

B

BACKFIRE — Ignition of the mixture in the intake manifold by flame from the cylinder such as might occur from a leaking intake valve.

BACKLASH — The clearance or "play" between two parts, such as meshed gears.

BACK-PRESSURE — A resistance to free flow, such as a restriction in the exhaust line.

BALL BEARING — An antifriction bearing consisting of a hardened inner and outer race with hardened steel balls interposed between the two races.

BEARING — A part in which a journal, pivot, or the like turns or moves.

BLOWBY — A leakage or loss of compression past the piston ring between the piston and the cylinder.

BOILING POINT — The temperature at which bubbles or vapors rise to the surface of a liquid and escape.

BORE — The diameter of a hole, such as a cylinder; also to enlarge a hole as distinguished from making a hole with a drill.

BREAK-IN — The process of wearing in to a desirable fit between the surfaces of two new or reconditioned parts.

BUSHING — A removable liner for a bearing.

BYPASS — An alternate path for a flow of air or liquid.

C

CALIBRATE — To determine or adjust the graduation or scale of any measuring instrument.

CAM-GROUND PISTON — A piston ground to a slightly oval shape which under the heat of operation becomes round.

CAMSHAFT — The shaft containing lobes or cams to operate the engine valves.

CARBON MONOXIDE — Gas formed by incomplete combustion. Colorless, odorless, and very poisonous.

CARBONIZE — The process of carbon formation within an engine, such as on the spark plugs and within the combustion chamber.

CARBURETOR — A device for automatically mixing gasoline fuel in the proper proportion with air to produce a combustible vapor.

CARBURETOR "ICING" — A term used to describe the formation of ice on a carburetor throttle plate during certain atmospheric conditions.

CETANE — Measure of ignition quality of diesel fuel — at what pressure and temperature the fuel will ignite and burn.

CHAMFER — A bevel or taper at the edge of a hole.

CHOKE — A device such as a valve placed in a carburetor air inlet to restrict the volume of air admitted.

COMBUSTION — The process of burning.

COMBUSTION CHAMBER — The volume of the cylinder above the piston with the piston on top center.

COMPRESSION — The reduction in volume or the "squeezing" of a gas. As applied to metal, such as a coil spring, compression is the opposite of tension.

COMPRESSION RATIO — The volume of the combustion chamber when the piston is at top-dead-center, as compared to the volume of the cylinder and chamber with the piston at bottom-dead-center.
Example: 8 to 1

COMPRESSION RELEASE MECHANISM — A device which opens a valve or port in the engine to slightly reduce the pressure in the combustion chamber for easier starting.

COMPRESSION STROKE — One of the four strokes of a 4-cycle engine.

CONDENSATION — The process of a vapor becoming a liquid; the reverse of evaporation.

CONNECTING ROD — Rod that connects the piston to the crankshaft.

CONTRACTION — A reduction in mass or dimension; the opposite of expansion.

CONTROL RACK — A part of an in-line diesel fuel injection pump which regulates the amount of fuel delivered to each cylinder.

CONVECTION — A transfer of heat by circulating heated air.

CORRODE — To eat away gradually as if by gnawing, especially by chemical action, such as rust.

COUNTERBORE — To enlarge a hole to a given depth.

COUNTERSINK — To cut or form a depression to allow the head of a screw to go below the surface.

COUNTERWEIGHT — Balancing weights in the crankshaft located opposite the crankpins.

CRANKCASE — The lower housing in which the crankshaft and many other parts of the engine operate. On a 2-cycle engine, the area in which the fuel-oil mixture is drawn before being transferred to the cylinder.

CRANKCASE DILUTION — When unburned fuel finds its way past the piston rings into the crankcase oil, where it dilutes or "thins" the engine lubricating oil.

CRANKCASE VENT — A device which allows the crankcase to breath, reducing pressures within the crankcase.

CRANKPIN — The bearing surface on the crankshaft around which the connecting rod is fastened. Also known as the connecting rod journal.

CRANKSHAFT — The main drive shaft of an engine which takes reciprocating motion and converts it to rotary motion.

CRANKSHAFT COUNTERBALANCE — A series of weights attached to or forged integrally with the crankshaft to offset the reciprocating weight of each piston and rod.

CRUDE OIL — Liquid oil as it comes from the ground.

CU. IN. — Cubic Inch

CYLINDER — A round hole having some depth bored to receive a piston; also sometimes referred to as "bore" or "barrel."

CYLINDER BLOCK — The largest single part of a 4-cycle engine. The basic or main mass of metal in which the cylinders are bored or placed. The part which supports other engine components.

CYLINDER HEAD — A detachable portion of an engine fastened securely to the cylinder block which contains all or a portion of the combustion chamber. Some 2-cycle engines contain cylinder heads which are not removable.

CYLINDER LINER — A sleeve or tube interposed between the piston and the cylinder wall or cylinder block to provide a readily renewable wearing surface for the cylinder.

D

DEAD CENTER — The extreme top or bottom position of the crankshaft throw at which the piston is not moving in either direction.

DELIVERY VALVE — A valve in a diesel fuel injection pump which distributes fuel in equal quantities to all cylinders.

DENSITY — Compactness; relative mass of matter in a given volume.

DETERGENT — A compound of a soap-like nature used in engine oil to remove engine deposits and hold them in suspension in the oil.

DETONATION — A too rapid burning or explosion of the mixture in the engine cylinders. It becomes audible through a vibration of the combustion chamber walls and is sometimes confused with a "ping" or spark "knock."

DIAGNOSIS — In engine service, the use of instruments to "troubleshoot" the engine parts to locate the cause of a failure.

DIESEL ENGINE — Named after its developer, Dr. Rudolph Diesel. This engine ignites fuel in the cylinder from the heat generated by compression. The fuel is an "oil" rather than gasoline, and no spark plug or carburetor is required. Instead, a fuel injection pump and injection nozzle are used.

DIESEL FUEL — A petroleum fuel prepared primarily for use as fuel for diesel engines.

DILUTION — See CRANKCASE DILUTION.

DISPLACEMENT — See PISTON DISPLACEMENT.

DISTORTION — A warpage or change in form from the original shape.

DOWEL PIN — A pin inserted in matching holes in two parts to maintain those parts in fixed relation one to the other.

DRAWBAR POWER — Measure of the pulling power of a machine at the drawbar hitch point.

DYNAMOMETER — A test unit for measuring the actual power produced by an engine.

E

ECCENTRIC — One circle within another circle but with a different center of rotation. An example of this is a driving cam on a camshaft.

ENERGY — The capacity for doing work.

ENGINE — The prime source of power generation used to propel the machine.

ENGINE DISPLACEMENT — The sum of the displacement of all the engine cylinders. See PISTON DISPLACEMENT.

EVAPORATION — The process of changing from a liquid to a vapor, such as boiling water to produce steam. Evaporation is the opposite of condensation.

EXHAUST MANIFOLD — The passages from the engine cylinders to the muffler which conduct the exhaust gases away from the engine.

EXHAUST STROKE — One of the four strokes of a 4-cycle engine.

EXPANSION — An increase in size. For example, when a metal rod is heated it increases in length and perhaps also in diameter. Expansion is the opposite of contraction.

EXTERNAL-GEAR PUMP — A type of oil pump used on 4-cycle engines. Oil, trapped between gear teeth and the housing, is pressurized when the gear teeth mesh.

F

FILTER — (Oil, Water, Gasoline, Etc.) — A unit containing an element, such as a screen of varying degrees of fineness. The screen of filtering element is made of various materials depending upon the size of the foreign particles to be eliminated from the fluid being filtered.

FIRING ORDER — On multiple-cylinder engines, the sequence in which the fuel-air mixture is ignited in the cylinders.

FLOATING PISTON PIN — A piston pin which is not locked in the connecting rod or the piston, but is free to turn or oscillate in both the connecting rod and the piston.

FLOAT LEVEL — The height of the fuel in the carburetor bowl, usually regulated by means of a suitable valve and float.

"FLUTTER" OR "BOUNCE" — In engine valves, refers to a condition where the valve is not held tightly on its seat during the time the cam is not lifting it.

FLYWHEEL — A heavy wheel in which energy is absorbed and stored by means of momentum. It is usually attached to one end of the engine crankshaft.

FOOT-POUND (Ft.-Lb.) — This is a measure of the amount of energy or work required to lift one pound a distance of one foot. The equivalent metric measure is the Newton Meter.

FOUR-STROKE-CYCLE ENGINE — An engine in which a fuel-air mixture is burned in the cylinder every other revolution of the crankshaft. These strokes are (1) intake stroke; (2) compression stroke; (3) power stroke; (4) exhaust stroke.

FUEL KNOCK — Same as Detonation.

FUEL PUMP — A mechanical or electric device used in force-feed fuel systems to transfer fuel from the fuel tank to the carburetor or fuel injection pump.

G

GAS — A substance which can be changed in volume and shape according to the temperature and pressure applied to it. For example, air is a gas which can be compressed into smaller volume and into any shape desired by pressure. It can also be expanded by the application of heat.

GASOLINE — A highly volatile fuel used in internal combustion engines.

GEAR RATIO — The number of revolutions made by a driving gear as compared to the number of revolutions made by a driven gear of different size. For example, if one gear makes three revolutions while the other gear makes one revolution, the gear ratio would be 3 to 1.

GLAZE — As used to describe the surface of the cylinder, an extremely smooth or glossy surface such as a cylinder wall highly polished over a long period of time by the friction of the piston pin.

GOVERNOR — A device to control and regulate speed. May be mechanical, hydraulic, or electrical.

GRIND — To finish or polish a surface by means of an abrasive wheel.

H

HEAT TREATMENT — A combination of heating and cooling operations timed and applied to a metal in a solid state in a way that will produce desired properties.

HOLE GAUGE — See MICROMETER (INSIDE MICROMETER).

HONE — An abrasive tool for correcting small irregularities or differences in diameter in a cylinder.

HORSEPOWER (HP) — The energy required to lift 550 lbs. one foot in one second. The equivalent metric unit of measure is the Kilowatt.

HOT SPOT — Refers to a comparatively thin section or area of the wall between the inlet and exhaust manifold of an engine, the purpose being to allow the hot exhaust gases to heat the comparatively cool incoming mixture. Also used to designate local areas of the cooling system which have attained above average temperatures.

I

I.D. — Inside diameter

IDLE — Refers to the engine operating at its slowest speed with a machine under no load.

INERTIA — A physical law which tends to keep a motionless body at rest or also tends to keep a moving body in motion; effort is thus required to start a mass moving or to retard or stop it once it is in motion.

I-HEAD ENGINE — A 4-cycle engine with the valves located directly above the piston.

INHIBITOR — A material to restrain some unwanted action, such as a rust inhibitor which is a chemical added to cooling systems to retard the formation of rust.

INJECTION NOZZLE (Diesel) — An assembly which receives a metered charge of fuel from another source at relatively low pressure, then is actuated to inject the charge of fuel into a cylinder or chamber at high pressure and at the proper time.

INJECTION PUMP (Diesel) — A device by means of which the fuel is metered and delivered under pressure to the injection nozzle.

IN-LINE CYLINDERS — A multiple-cylinder engine configuration where all cylinders are located in a straight line.

INPUT SHAFT — The shaft carrying the driving gear, such as in a transmission by which the power is applied.

INTAKE MANIFOLD — The passages which conduct the fuel-air mixture from the carburetor to the engine cylinders.

INTAKE STROKE — One of the four strokes of a 4-cycle engine.

INTAKE VALVE — A valve which permits a fluid or gas to enter a chamber and seals against exit.

INTEGRAL — The whole made up of parts.

INTERNAL COMBUSTION — The burning of a fuel within an enclosed space.

J

JET — A small orifice in the carburetor through which fuel passes.

JOURNAL — A part or support within which a shaft operates.

K

KEY — A small block inserted between the shaft and hub to prevent circumferential movement.

KEYWAY OR KEYSEAT — A groove or slot cut for inserting a key to hold a part on a shaft, etc.

KILOPASCALS — The metric measurement of pressure. The equivalent customary measurement is pounds per square inch or PSI.

KILOWATTS — The energy required to move a force of 1000 newtons a distance of one meter in one second.

KNOCK — A general term used to describe various noises occurring in an engine; may be used to describe noises made by loose or worn mechanical parts, preignition, detonation, etc.

KNURLED — Displacing metal on the skirt of a piston to increase the diameter of the piston (on automotive engines).

L

LACQUER — A solution of solids in solvents which evaporate with great rapidity.

LAPPING — The process of fitting one surface to another by rubbing them together with an abrasive material between the two surfaces.

L-HEAD ENGINE — An engine design in which both valves are located on one side of the engine cylinder.

LINER — Usually a thin section placed between two parts, such as a replaceable cylinder liner in an engine.

LIQUID COOLING — A method of cooling an engine where a liquid is circulated through the engine water jacket, absorbs heat, and is circulated back to the radiator where some of the heat is dissipated.

M

MANIFOLD — A pipe or casting with multiple openings used to connect various cylinders to one inlet or outlet.

MICROMETER — A measuring instrument for either external or internal (inside micrometer) measurement in thousandths and sometimes tenths of thousandths of inches.

MISFIRING — Failure of an explosion to occur in one or more cylinders while the engine is running; may be a continuous or intermittent failure.

MOTOR — This term should be used in connection with an electric motor and should not be used when referring to the engine of a machine.

MUFFLER — A chamber attached to the end of the exhaust pipe which allows the exhaust gases to expand and cool. It is usually fitted with baffles or porous plates and serves to subdue much of the noise created by the exhaust.

N

NEEDLE BEARING — An antifriction bearing using a great number of long, thin rollers.

NEWTON — The force required to move a mass of one kilogram a speed of one meter per second squared.

NEWTON METER — The metric measurement of the amount of energy or work required to lift one newton of force a distance of one meter.

O

OCTANE — Measurement which indicates the tendency of gasoline to detonate or knock.

O.D. — Outside diameter

OIL PUMPING — A term used to describe an engine which is using an excessive amount of lubrication oil because the oil escapes past the piston rings into the combustion chamber.

OPPOSED CYLINDERS — A multiple-cylinder engine configuration where the cylinders are located opposite of each other.

P

PEEN — To stretch or clinch over by pounding with the rounded end of a hammer.

PETROLEUM — A group of liquid and gaseous compounds composed of carbon and hydrogen which are removed from the earth.

PINION — A small gear having the teeth formed on the hub.

PISTON — A cylindrical part closed at one end which is connected to the crankshaft by the connecting rod. The force of the expansion in the cylinder is exerted against the closed end of the piston, causing the connecting rod to move the crankshaft.

PISTON DISPLACEMENT — The volume of air moved or displaced by moving the piston from one end of its stroke to the other.

PISTON HEAD — That part of the piston above the rings.

PISTON LANDS — Those parts of a piston between the piston rings.

PISTON PIN — A cylinder which fits through both the piston and the connecting rod to connect both parts.

PISTON RING — An expanding ring placed in the grooves of the piston to seal off the passage of fluid or gas past the piston.

PISTON RING EXPANDER — A spring placed behind the piston in the groove to increase the pressure of the ring against the cylinder wall. Also, the name of a tool used to remove and install piston rings.

PISTON RING GAP — The clearance between the ends of the piston ring.

PISTON RING GROOVE — The channel or slots in the piston in which the piston rings are placed.

PISTON RING SIDE CLEARANCE — The clearance between the side of the piston ring and its groove.

PISTON SKIRT — That part of the piston below the rings.

PLUNGER PUMP — A type of oil pump used on 4-cycle engines. It contains a plunger which moves up and down inside a barrel.

PORT — The openings in the cylinder block for valves, exhaust and inlet pipes, or water connections. In two-cycle engines, the openings for intake and exhaust.

POWER STROKE — One of the four strokes of a 4-cycle engine.

PREIGNITION — Ignition occurring earlier than intended. For example, the explosive mixture being fired in a cylinder as by a flake of incandescent carbon before the electric spark occurs.

PRESS-FIT — Also known as a force-fit or drive-fit. This term is used when the shaft is slightly larger than the hole and must be forced into place.

PRESSURE-TYPE LUBRICATION — A lubrication system which uses a pump to pressurize oil to be delivered to engine components.

PSI — A measurement of pressure in pounds per square inch. The equivalent metric measurement is kilopascal.

PUSH ROD — A connecting link in an operating mechanism, such as the rod interposed between the valve lifter and rocker arm on an overhead valve engine.

R

RACE — As used with reference to bearings; a finished inner and outer surface in which or on which balls or rollers operate.

RADIATOR — A part of the liquid cooling system through which hot water passes for cooling.

RATED POWER — Valve used by the engine manufacturer to rate the power of his engine, allowing for safe loads, etc.

RATIO — The relation or proportion of one number or quantity to another.

REAM — To finish a hole accurately with a rotating fluted tool.

RECIPROCATING MOTION — A back and forth movement, such as the action of a piston in a cylinder.

RECOIL STARTER — A mechanism used on some engines which requires the operator to pull a starter rope to start the engine.

REED VALVE — A valve used in the crankcase of some 2-cycle engines to admit the fuel-oil mixture to the engine.

RIDGE REAMING — The process of removing a ridge of metal from the top of the cylinder bore. The ridge is formed when the piston reciprocates and wears away some metal from the rest of the cylinder bore.

ROCKER ARM — In an engine a lever located on a fulcrum or shaft, one end of the valve stem, the other on the push rod.

ROLLER BEARING — An inner and outer race upon which hardened steel rollers operate.

ROTOR PUMP — A type of oil pump used on 4-cycle engines. An inner rotor rotates inside an outer rotor and pressurizes the oil as it rotates.

ROTARY MOTION — A circular movement, such as the rotation of a crankshaft.

RPM — Revolutions per minute

RUNNING-FIT — Where sufficient clearance has been allowed between the shaft and journal to allow free running without overheating.

S

SAND BLAST — To clean a surface by means of sand propelled by compressed air.

SCALE — A flaky deposit occurring on steel or iron. Ordinarily used to describe the accumulation of minerals and metals accumulating in an engine cooling system.

SCAVENGING — A method of removing burned gases from the combustion chamber and replacing the gases with a new charge of fuel-air mixture.

SCORE — A scratch, ridge or groove marring a finished surface.

SEAT — A surface, usually machined, upon which another part rests or seats. For example, the surface upon which a valve face rests.

SHIM — Thin sheets used as spacers between two parts, such as the two halves of a journal bearing.

SHRINK-FIT — Where the shaft or parts are slightly larger than the hole in which it is to be inserted. The outer part is heated above its normal operating temperature or the inner part chilled below its normal operating temperature and assembled in this condition; upon cooling an exceptionally tight fit is obtained.

SLIDING-FIT — Where sufficient clearance has been allowed between the shaft and journal to allow free running without overheating.

SLUDGE — A composition of oxidized petroleum products along with an emulsion formed by the mixture of oil and water. This forms a pasty substance and clogs oil lines and passages and interferes with engine lubrication.

SOLID INJECTION — The system used in diesel engines where fuel as a fluid is injected into the cylinder rather than a mixture of fuel and air.

SOLVENT — A solution which dissolves some other material. For example, water in a solvent for sugar.

SPLASH-TYPE LUBRICATION — A lubrication system which uses a dipper or a slinger to splash oil onto engine components.

SPLINE — A long keyway.

STRESS — The force or strain to which a material is subjected.

STROKE — The distance a piston moves.

STUDS — A rod with threads cut on both ends, such as a cylinder stud which screws into the cylinder block on one end and has a nut placed on the other end to hold the cylinder head in place.

SUCTION — Suction exists in a vessel when the pressure is lower than the atmospheric pressure; also see VACUUM.

SYNCHRONIZE — To cause two events to occur at the same time. For example, to time a mechanism so that two or more sparks will occur at the same instant.

T

TACHOMETER — A device for measuring and showing the rotating speed of an engine. Also see VIBRATION TACHOMETER.

TAP — To cut threads in a hole with a tapered, fluted, threaded tool.

TAPPET — The adjusting device for varying the clearance between the valve stem and the cam. May be built into the valve lifter in an engine or may be installed in the rocker arm or an overhead valve engine.

T.D.C. — Top dead center (of a piston).

TELESCOPING GAUGE — See MICROMETER (Inside Micrometer).

THERMOSTAT — A heat controlled valve used in the cooling system of an engine to regulate the flow of water between the cylinder block and the radiator.

THROTTLE — A valve which controls the amount of air and fuel flow through the carburetor.

THROW — The distance from the center of the crankshaft main bearing to the center of the crankpin (connecting rod journal).

TIMING GEARS — Any group of gears which are driven from the engine crankshaft to cause the valves, ignition and other engine driven accessories to operate at the desired time during the engine cycle.

TOLERANCE — A permissible variation between the two extremes of a specification of dimensions. Used in the precision fitting of mechanical parts.

TORQUE — The effort of twisting or turning.

TORQUE WRENCH — A special wrench with a built in indicator to measure the applied turning force.

TRANSFER PUMP — See FUEL PUMP.

TROUBLESHOOTING — A process of diagnosing or locating the source of the trouble or troubles from observation and testing. Also see DIAGNOSIS.

TUNE-UP — A process of accurate and careful adjustments to obtain the best engine performance.

TURBINE — A series of angled blades located on a wheel against which fluids or gases are impelled to rotate a shaft. Part of a turbocharger.

TURBOCHARGER — An exhaust-driven mechanism which forces greater quantities of air into the engine so more fuel can be mixed with the air and more power generated.

TURBULENCE — A disturbed, irregular motion of fluids or gases.

TWO-CYCLE ENGINE — An engine design permitting a power stroke once for each revolution of the crankshaft.

V

VACUUM — A perfect vacuum has not been created as this would involve an absolute lack of pressure. The term is ordinarily used to describe a partial vacuum; that is, a pressure less that atmospheric pressure; in other words a depression.

VACUUM GAUGE — An instrument designed to measure the degree of vacuum existing in a chamber.

VALVE (INTAKE AND EXHAUST) — A device for opening and sealing the cylinder intake and exhaust ports.

VALVE CLEARANCE — The air gap allowed between the end of the valve stem and the valve lifter or rocker arm to assure the valve closes completely when it is seated.

VALVE FACE — That part of a valve which mates with and rests upon a seating surface.

VALVE GRINDING — Also called valve lapping. A process of lapping or mating the valve seat and valve face usually performed with the air of an abrasive.

VALVE HEAD — The portion of the valve upon which the valve face is machined.

VALVE-IN-HEAD ENGINE — An engine (4-cycle) in which the valves are located in the cylinder head.

VALVE LIFTER — A push rod or plunger placed between the cam and the valve on an engine; is often adjustable to vary the length of the unit.

VALVE MARGIN — The space or rim between the surface of the head and the surface of the valve face.

VALVE SEAT — The matched surface upon which the valve face rests when the valve is closed.

VALVE SPRING — A spring attached to a valve to return it to the seat after it has been released from the lifting or opening means.

VALVE STEM — That portion of a valve which rests within a guide.

VALVE STEM GUIDE — A bushing or hole in which the valve stem is placed which allows lateral motion only.

VALVE TRAIN — All of the parts in a 4-cycle engine that work together to operate the valves.

VANES — Any plate, blade or the like attached to an axis and moved by or in air or a liquid.

VAPORIZER — A device for transforming or helping to transform a liquid into a vapor; often includes the application of heat.

VAPOR LOCK — A condition wherein the fuel boils in the fuel system forming bubbles which retard or stop the flow of fuel to the carburetor.

VENTURI — Two tapering streamlined tubes joined at their small ends so as to reduce the internal diameter.

VIBRATION TACHOMETER — A tool which senses engine vibrations to determine the speed at which the engine is running.

VISCOSITY — The resistance to flow of an oil.

VOLATILITY — The tendency for a fluid to evaporate rapidly or pass off in the form of vapor. For example, gasoline is more volatile than kerosene as it evaporates at a lower temperature.

W

WATER PUMP — A part of the liquid-cooling system which pumps coolant through the engine.

WINDUP STARTER — A mechanism used on some engines which requires the operator to turn a hand crank in the process of starting the engine.

INDEX